Palgrave Studies in Literature, Science and Medicine

Palgrave Studies in Literature, Science and Medicine is an exciting new series that focuses on one of the most vibrant and interdisciplinary areas in literary studies: the intersection of literature, science and medicine. Comprised of academic monographs, essay collections, and Palgrave Pivot books, the series will emphasize a historical approach to its subjects, in conjunction with a range of other theoretical approaches. The series will cover all aspects of this rich and varied field and is open to new and emerging topics as well as established ones.

More information about this series at
http://www.palgrave.com/gp/series/14613

Natalie Roxburgh · Jennifer S. Henke
Editors

Psychopharmacology in British Literature and Culture, 1780–1900

palgrave
macmillan

Editors
Natalie Roxburgh
University of Siegen
Siegen, Germany

Jennifer S. Henke
University of Bremen
Bremen, Germany

Palgrave Studies in Literature, Science and Medicine
ISBN 978-3-030-53597-1 ISBN 978-3-030-53598-8 (eBook)
https://doi.org/10.1007/978-3-030-53598-8

Cover credit: Nathaniel Noir/Alamy Stock Photo

This Palgrave Macmillan imprint is published by the registered company Springer Nature Switzerland AG
The registered company address is: Gewerbestrasse 11, 6330 Cham, Switzerland

Praise for *Psychopharmacology in British Literature and Culture, 1780–1900*

"This pioneering study of drug effects, not just addiction, in the nineteenth century ranges from opium to alcohol, lavender water, wormwood, and other herbal substances. Roxburgh and Henke have done a service for the fields of both medical humanities and literature and science by revealing the important role literary and cultural texts played in making possible the emergence of psychopharmacology in the next century."

—Jay Clayton, *William R. Kenan Professor and Director of the Curb Center, Vanderbilt University, USA*

"This collection offers useful consideration of the history of late-eighteenth- and nineteenth-century pharmacology; its essays suggest all kinds of possibilities for new conversations about medical science and the literary imagination."

—Adam Colman, *author of* Drugs and the Addiction Aesthetic in Nineteenth-Century Literature *(Palgrave Macmillan, 2019)*

Contents

1 **Situating Psychopharmacology in Literature and Culture** 1
Natalie Roxburgh and Jennifer S. Henke

Part I Drugs and Genre

2 **Historicising Keats' Opium Imagery Through Neoclassical Medical and Literary Discourses** 23
Octavia Cox

3 **"Grief's Comforter, Joy's Guardian, Good King Poppy!": Opium and Victorian Poetry** 47
Irmtraud Huber

4 **Dangerous Literary Substances: Discourses of Drugs and Dependence in the Nineteenth-Century Sensation Novel Debates** 69
Sarah Frühwirth

Part II Rethinking the Pharmacological Body: Drugs and the Borders of the Human

5 **Blurring Plant and Human Boundaries: Erasmus Darwin's *The Loves of the Plants*** 93
C. A. Vaughn Cross

6 **Pharmacokinetics and Opium-Eating: Metabolites, Stomach Aches and the Afterlife of De Quincey's Addiction** 117
Hannah Markley

7 **A Posthumanist Approach to Agency in De Quincey's *Confessions*** 135
Anna Rowntree

Part III The Cultural Politics of Known Drug Effects

8 **Reading De Quinceyan Rhetoric Against the Grain: An Actor-Network-Theory Approach** 155
Anuj Gupta

9 **Blood Streams, Cash Flows and Circulations of Desire: Psychopharmacological Knowledge About Opium in Nineteenth-Century Women's Fiction** 171
Nadine Böhm-Schnitker

10 **The Indeterminate Pharmacology of Absinthe in Nineteenth-Century Literature and Beyond** 195
Vanessa Herrmann

Part IV Historicising the Prescription: Medication and Self-Medication

11 "She Furnishes the Fan and the Lavender Water": Nervous Distress, Female Healers and Jane Austen's Herbal Medicine 217
Rebecca Spear

12 "When Poor Mama Long Restless Lies, / She Drinks the Poppy's Juice": Opium and Gender in British Romantic Literature 241
Joseph Crawford

13 *Middlemarch* and Medical Practice in the Regency Era: From "Bottles of Stuff" to the Clinical Gaze 263
Björn Bosserhoff

Index 287

Notes on Contributors

Nadine Böhm-Schnitker (PD Dr.) is an Assistant Professor (*Oberassistentin*) in British and American Studies at the University of Bielefeld. She specialises in Victorian and Neo-Victorian Studies and has recently completed the manuscript for a monograph titled *Senses and Sensations: Towards an Aesthetics of the Victorian Novel*. Apart from researching hunger, her current projects deal with the cultural legacy of the Opium Wars, neo-Victorian negotiations of class and inheritance and practices of comparison in the long eighteenth century.

Björn Bosserhoff (Ph.D.) is a Research Associate at the University of Bonn's North American Studies Program as well as a freelance academic editor and translator. He is the author of *Radical Contra-Diction: Coleridge, Revolution, Apostasy* (Cambridge Scholars, 2016), a book that re-evaluates S.T. Coleridge's political trajectory from 'radical' to 'conservative'. With Sabine Sielke, he has co-edited the volumes *Nostalgia: Imagined Time-Spaces in Global Media Cultures* (Peter Lang, 2016) and *New York, New York! Urban Spaces, Dreamscapes, Contested Territories* (Peter Lang, 2015). His research interests include the social and cultural history of the British long nineteenth century, American history and politics, art history, religious history, the history o f psychiatry, addiction theory, the history of feminism and queer studies.

Octavia Cox received her doctorate from the University of Oxford, and she now teaches and researches at the University of Nottingham and

University of Oxford. She has published on women's writing in *The Lady's Poetical Magazine (1781–82)* and on post-colonial implications of the Reverse-Robinsonade genre in *The Woman of Colour* (1808). She is currently writing her first monograph, *Alexander Pope in the Romantic Age*.

Joseph Crawford is a Senior Lecturer in English Literature at the University of Exeter. He is the author of four books: *Raising Milton's Ghost* (2011), *Gothic Fiction and the Invention of Terrorism* (2013), *The Twilight of the Gothic* (2014) and *Inspiration and Insanity in British Poetry* (2019). His current research deals with the literature of the British post-Romantic period.

Sarah Frühwirth is a Ph.D. candidate at the University of Vienna, Austria. In 2015, she received a prestigious three-year research fellowship for her Ph.D. project "Free Will and Determinism in the British Sensation Novel of the 1860s and 1870s", which investigates to what extent the sensation novelists' depiction of free will and determinism was influenced by contemporary scientific and philosophical discourses. In 2018, she became the reviews editor of *The Wilkie Collins Journal*. Her research interests include sensation fiction, nineteenth-century fiction in general, the history of medicine and science and food studies.

Anuj Gupta is a Ph.D. Candidate in Rhetoric and Composition at the University of Arizona (USA) as of 2020. Gupta took his M.Phil. in English Literature and Cultural Studies from the Centre of English Studies at the Jawaharlal Nehru University and was Assistant Director of the Young India Fellowship's Critical Writing programme at Ashoka University (both in India). His work emphasises the way rhetoric shapes our understanding of science and sociocultural phenomena.

Dr. Jennifer S. Henke is an Assistant Professor at the University of Bremen, Germany. She was awarded a Ph.D. in Anglophone literature and culture with a book on gender and space in contemporary Shakespeare films (*Unsex Me Here*, Wissenschaftlicher Verlag Trier 2014). Since 2013, she has been a member of the interdisciplinary and international research group *Fiction Meets Science* in cooperation with the Universities of Hamburg, Oldenburg, Bielefeld, Sydney, San Luis Potosi and Guelph. Her research and teaching activities include literatures of addiction, science and the media, gender and mathematics, Irish politics, medical humanities, body studies, gender and science-fiction, posthumanism as well as *Frankenstein* in literature and film. She is currently working on her second book that deals with medicine and the pregnant female body in eighteenth-century literature and culture.

Vanessa Herrmann holds two MAs (2004 and 2019) and a Ph.D. (2011) from the University of Siegen and is currently a Lecturer at the University of Bremen. Her dissertation *Shakespeare Reloaded: The Shakespeare Renaissance 1989–2004* was published in 2011 (Wissenschaftlicher Verlag Trier). Her most recent publications include "Adaptation as City Branding: The Case of *Dexter* and Miami" in the collection *Where is Adaptation? Mapping Cultures, Texts, and Contexts* (Eds. Casie Hermansson and Janet Zepernick; John Benjamins Publishing Company, 2018). Her research interests are adaptation and game studies as well as Victorian literature and culture.

Irmtraud Huber is a Lecturer in the English Department of the Ludwig-Maximilians University of Munich and an Associated Postdoctoral Fellow of the Walter Benjamin Kolleg, University of Bern. With support from the Swiss National Science Foundation, she has held Visiting Fellowships at Columbia University, at the University of Cambridge and at Queen Mary College, University of London. She has published two monographs on contemporary fiction, *Literature After Postmodernism* (Palgrave, 2014) and *Present-tense Narration in Contemporary Fiction* (Palgrave, 2016). She has also guest-edited a special edition of *EJES* on *Poetry, Science and Technology* with Wolfgang Funk. Her research interests include time in literature, the epistemology of literary form and the theory of lyric. Her current research focuses on (re)configurations of time in Victorian poetry.

Hannah Markley is a Marion L. Brittain Postdoctoral Fellow at the Georgia Institute of Technology. Her work has appeared in *European Romantic Review, Essays in Romanticism* (Taylor & Francis) and *Parallax* (Taylor & Francis). Her current book project takes as its point of departure the nineteenth-century fascination with sisters as mourners to show how authors of the period use female bodies to construct gendered visions of survival in an ecologically precarious world.

Anna Rowntree (M.A.) is a freelance writer and yoga teacher based in Berlin, Germany.

Natalie Roxburgh (Ph.D. 2011, Rutgers) is a Lecturer and Research Fellow in English Literary Studies at the University of Siegen in Germany. Her current book-in-progress is on rethinking aesthetic disinterestedness in nineteenth-century Britain. She has published widely on a variety of topics—such as science, economics and politics—from the seventeenth century to the present, including a monograph titled *Representing Public Credit: Credible Commitment, Fiction, and the Rise of the Financial Subject* (Routledge, 2016). Her work can be read in *Eighteenth-Century Fiction*, *Mosaic* and many other places. She currently has three essays out on the rise of psychopharmacology in contemporary fiction and film, two of which are published by Palgrave.

Rebecca Spear is completing her Ph.D. at Cardiff University. Her thesis explores Jane Austen's interactions with science, focusing on medicine and botany. She has a B.A. (Hons) from the University of South Wales and an M.A. from Cardiff University, both in English Literature. Since 2015, she has been an editorial assistant for the *Journal of Literature and Science*. Her research interests are women's writing of the long eighteenth century, literature and science, Romanticism and the history of psychiatry.

C. A. Vaughn Cross (Ph.D. in History, Auburn University) is an Assistant Professor in the Department of Geography and Sociology at Samford University where she teaches interdisciplinary courses involving the history of technology and culture, religion, sex and gender. She has published articles in *Social Sciences & Missions*, *The Indian Journal of American Studies* and *Perspectives in Religious Studies* as well as in the anthologies *North American Foreign Missions, 1810–1914: Theology, Theory, and Policy* (Ed. Wilbert R Shenk; William B. Eerdmans Publishing Company, 2004) and *Transhumanism and the Church: Chips in the Brain, Immortality, and the World of Tomorrow* (Eds. Steve Donaldson and Ron Cole-Turner; Palgrave Macmillan, 2018).

CHAPTER 1

Situating Psychopharmacology in Literature and Culture

Natalie Roxburgh and Jennifer S. Henke

In the last decades, scholars have become increasingly interested in the role psychoactive substances play in the making of and expression of human culture. Given the proliferation of documentaries, news items and political debates on decriminalisation in recent years, it is perhaps no wonder that drugs have become a focal point of scholarly concern. Indeed, in the public spotlight are issues such as the off-label (over-)use of medication, the proliferation of opioid addiction (the 'opioid epidemic') in the last decade through over-prescription in the United States and a scientific and cultural reassessment of real risks that both legal and illegal drugs pose. Such topics, despite their more recent political, social and

N. Roxburgh (✉)
University of Siegen, Siegen, Germany
e-mail: natalie.roxburgh@uni-siegen.de

J. S. Henke
University of Bremen, Bremen, Germany
e-mail: j.henke@uni-bremen.de

N. Roxburgh and J. S. Henke (eds.), *Psychopharmacology in British Literature and Culture, 1780–1900*, Palgrave Studies in Literature, Science and Medicine,
https://doi.org/10.1007/978-3-030-53598-8_1

cultural resonances, have been salient for centuries, at least insofar as they appear in literary and cultural texts.[1]

This volume historicises the way medical and scientific knowledge came to provide systematic accounts of how drugs work by honing the way they are represented in literary and cultural texts, which challenge, anticipate, interrogate, participate in and criticise their medical counterparts. Most studies on drugs in literature and culture have focused on the history of addiction, and many have used literary biography as the main source texts (Milligan 1995, 2005; Davenport-Hines 2001; Boon 2002; Redfield and Brodie 2002; Ronell 2004; Reed 2006; Zieger 2008; Jay 2011; Comitini 2012; Mangiavellano 2013; Foxcroft 2016; Malek), often focusing on Thomas De Quincey (Abrams 1971; Schiller 1976; Rzepka 1991; Clej 1995; Morrison and Roberts 2008; Morrison 2011). There is, however, much more to be said about psychoactive substances and the connections human beings have to them. As Susan Zieger points out, the expansion of international trade meant that people "became enchanted with marvelous substances from exotic locales: spices, sugar, tobacco, chocolate, coffee, tea, rum. Imperial commerce in the period from 1500 to 1800 laid the groundwork for a 'psychoactive revolution'" (4). During this period, and especially in the eighteenth and nineteenth centuries, knowledge about how drugs work on the brain, nerves and the body increased, and with this knowledge came a process of identifying and restricting their use. Breaking also from a tendency to emphasise the way writers used drugs as reflected in literary biography, this volume examines what contemporaries knew about how drugs affect the body, and what effects they have on mood, sensation, thinking and behaviour, in order to contribute to the discourse on addiction as well as to consider the cultural significance of psychoactive substances beyond addiction. There were, after all, many ways to use substances that were not based on drug-induced need.

The nineteenth century is a fascinating time to study drugs precisely because of the convergence of different medicines. One need only reflect on the experience of chemist Humphry Davy, who records his experimentation with nitrous oxide, stumbling his way into his own notion of the substance's effects and its subsequent use: "My labours are finished for the season as to public experimenting and enunciation. My last lecture was on Saturday evening. Nearly 500 persons attended, and amongst

[1] We would like to thank Norbert Schaffeld, Imke Grothenn and the Bremen English Studies Colloquium for support and feedback on this project.

other philosophers, your countryman, Professor Pictet. There was respiration, nitrous oxide, and unbounded applause—Amen. To-morrow a party of philosophers meet at the Institution, to inhale the joy-inspiring gas. It has produced a great sensation" (Davy 1858, 64). Experimentation with nitrous oxide (and a careful observation of its effects) led to the conclusion that the gas could function as an anesthetic drug: "Does not sensibility more immediately depend on respiration? [...] As nitrous oxide in its extensive operation appears capable of destroying physical pain, it may probably be used with advantage during surgical operations in which no great effusion of blood takes place" (Davy 1858, 18). Indeed, as Davy's experimentation shows, the nineteenth century was rife with attempts to ascertain particular effects of particular substances, which were put to use in an increasingly systematised way.

Coined by the pharmacologist David Macht in 1920, the term *psychopharmacology* is usually associated with the scientific study of drugs and their capacity to treat mental disorders. In this volume, we use the term to discuss the representation of drugs in late-eighteenth- and nineteenth-century literary and cultural contexts: contemporaries hypothesised about what might now be called *drug action* (even though in this time, drugs were thought to affect the body) and explored utilising certain substances for particular known *drug effects*. Our book differentiates itself from what has already been published on drugs in literature and culture by considering the role emergent psychopharmacological *knowledges* play in literary and cultural texts during the period when the field slowly began to develop. Nineteenth-century science was growing, dynamic and controversial—and with no separate concepts of psychology or psychiatry to speak of. The history of known drug effects dovetails with the development of others fields. The nineteenth century is crucial for the history of psychiatric medicine, for the development of a theory o f addiction and also for a theory of drug effect that moves beyond humouralism.

Abandoning humouralist theories of the body that posited nerve system as tubes of liquids (which left little room for a theory of drug effect), these new methods included new materialist medical theories that afforded mechanical agency to inert substances. Many embraced an organicist materialist view of the body (see Ruston 2012, 24). One such theory was that of Brunonianism, which scholars such as Gavin Budge and Roy Porter have emphasised were influential on the development of medicine in general and psychiatry in particular (Budge 2013, 12–13, 56–57; Porter 1988, 89). Scottish physician John Brown posited that afflictions were the body's nervous reaction to external stimuli caused by an under- or over-stimulation, dividing diseases into two classes, asthenic and sthenic, respectively. Drugs, and especially opium, were considered to be stimulants used to bring users back to a healthy state of equilibrium. Owing to Brown's influence on figures such as Thomas Beddoes (who, in turn, was a great influence on Samuel Taylor Coleridge)—as well as to the Opium Wars (1839–1860)—opium takes up a lot of space in this volume as it takes centre stage in the development of the science of drug effects. At the same time, alcohol, tea, tobacco and other herbal substances became increasingly relevant for medical knowledge. While the focus is on Britain, several of our essays show how Britain is interwoven with colonial contexts. Several of our essays address the impact of the Opium Wars, revealing a global circulation of drugs as well as colonial contexts for the known effects that go with them.

Studying psychopharmacology in the context of eighteenth- and nineteenth-century literature encompasses three areas: (1) considering what was known about the human brain, nerves and body, (2) accounting for contemporary knowledges about substances (usually plant-based) and (3) studying the manner in which literary texts represent how the use of drugs is embedded in specific cultural contexts. In what ways were drugs seen as empowering, healing or detrimental? What were the cultural—or even aesthetic—contexts for this assessment? How do these factors inflect the way substances are experienced? We address these questions in our four sections. First, our contributors examine the question of the aesthetic by looking at the relationship between genre and drugs. Second, we consider the way psychopharmacology in its cultural contexts puts pressure on a strict division between humans and drugs, and also humans and plants. Third, we assess the cultural and political influences that inform the way that known drug effects were described, discussed, classified and put to use. Finally, we conclude by examining early frameworks and attitudes towards medication and self-medication, thinking about the emergence of the prescription in literary texts.

Psychopharmacology in British Literature and Culture asks the question about what recent science and medicine say about drugs, and it historicises this knowledge with what contemporaries knew and thought about them. People take drugs for a variety of reasons: for medicine, for pleasure, for relieving withdrawal symptoms or for enhancing certain qualities for particular social or cultural ends. That is to say, the relationship between how drugs work on the brain (drug action) and how people think and feel while taking drugs (drug effects) is complex when we consider cultural contexts. And what we learn from literature cannot be deduced from a medical textbook alone. This book aims at extending the socio-historical investigations that concentrate on the discovery of drugs in the context of the history of psychopharmacology, on the boundaries between use and abuse in society, on the distribution and sale of (medicinal) drugs, on the drug trade and drug wars and on the way in which economic considerations have affected the determination of 'good', 'bad' or 'forbidden' drugs (Porter and Teich 1995; Drews 2000; Courtwright 2001; Curth 2006; Wallis 2012; Barbara 2015). Drugs are, first and foremost, material objects with specific chemical attributes that have particular benefits and/or detrimental effects depending on specific contexts. Our focus is on the drug itself, in whatever form the substance takes, an approach that factors in the way these chemicals were thought to work on the brain and/or body. Despite this volume's general focus on opium, examining several drugs in depth allows us to see that different cultural imaginaries of different drugs inform the way that they are interpreted. In so doing, articles in this collection add to the discussion by considering the way drug effects were thought to enable or enhance certain mental states and functions in various cultural contexts. In the end, this volume tells a story about the way a knowing use of drugs helped users become fit, or unfit, for certain social and cultural aims. Certainly, the early science of other drugs not emphasised in this volume—such as cocaine, which (as one gleans from the *oeuvre* of Arthur Conan Doyle) became popular at the end of the century when our volume closes—can also be better understood through the story we are telling (see Small 2016).

Readers will discover that essays here are located at the intersections of medical humanities and literature and science. Studies in the medical humanities have considered both the objective knowledge about the brain and body and the subjective experiences of being ill as well as changing discourses on madness and an ethics of prescribing drugs (Weber 2006; Petryna et al. 2006; Racine and Forlini 2010; Franke et al. 2015; Maier and Schaub 2015; Svenaeus 2017; Malleck 2020). In literature and science, interest in neuroscience and neurology (Stiles 2007; Walezak 2018; Servitje 2018) has been sparked by questions regarding the relationship between the brain (as part of the body) and the mind, the latter of which is represented in literature. Similarly, research in botany has provoked new understandings of the way plants were classified, used (by whom and with what social, geopolitical or colonial effects) and conveyed in literature (Schiebinger 2005, 2007, 2017; Campbell 2007; Martin 2011; Francia and Stobart 2015). Finally, research on drugs as chemical substances has explored the way people in the eighteenth and nineteenth centuries thought about and understood the relationship between the material brain and the subjective experiences attached to it, whether drugs were and are taken owing to addiction, for medicine, for pleasure or for spiritual reasons (Vice et al. 1994; Partridge 2018). The essays in this volume attempt to weave some of these threads together by starting with the drug substance itself. Our project is not to comprehensively study all drugs used in the late eighteenth and nineteenth centuries; rather, we examine the way knowledge about drugs in general is woven into cultural texts. Alcohol, tea, various herbs and wormwood are some of the drugs explored here in addition to opium.

In literary studies, a lot of attention has been given to the concept of the *pharmakon* made especially relevant by Jacques Derrida in his 1981 reading of Plato's *Phaedrus* in *Disseminations*. Meaning *poison* as well as *remedy*, at first glance the term seems contradictory (Derrida 1981, 97). The question of 'what is a *pharmakon*'—a remedy or poison—is related to the role of science in society, a sit is about identifying 'good' versus 'bad' drugs (see Herlinghaus 2018). After all, it is medical science that creates a taxonomy of illnesses and remedies, which in turn has cultural values and practices connected to them. This is something that literary texts capture, as they seek to represent the experience of drug-taking with science and cultural contexts attached, a task that medical literatures or science textbooks are not able to undertake by themselves. Our study distinguishes itself from the Derridean—and poststructuralist—tradition by taking the science of drugs quite literally: the *pharmakon* is, in the true sense of the word, a drug that can be a poison or a remedy, and it is not merely a stand-in for technology writ large (cf. Pies 2006; Stiegler 2011; Jenkins 2011; De Boever 2013). Another aspect that sets our volume apart from others is the consideration of the blurred boundaries between not only poison and remedy but also the drug as corrector and enhancer. It is pharmacological knowledge that helps to sort drugs into categories, which in turn have cultural influences and ramifications.

Late-eighteenth- and nineteenth-century literary texts analysed here do not all explicitly deal with the discourses of modern psychopharmacology, as this field is still very much in development. And yet, each essay explores the effects of drugs on the human brain and body in a way that would have been seen as adding to knowledge in a period in which psychopharmacology was still discovering itself: literary texts both anticipate and produce what will come next. Thus, this collection reads late-eighteenth- and nineteenth-century literature through what we now know about psychopharmacology while also rethinking the relevance of these literary texts for the development of the field.

Drugs and Genre

In some cases, a drug's established psychopharmacological effects were historically linked to how texts in various genres explained to the reading public how these effects come about. Although various plants were explored and represented in literature from the seventeenth to the nineteenth century, the most prevalent drug given literary treatment was opium, although alcohol's effects also garnered attention. Octavia Cox's "Historicising Keats' Opium Imagery through Neoclassical Medical and Literary Discourses" sets the stage for our volume by tracing the way opium was represented over the course of several centuries by focusing on two distinct genres: medical texts and literary ones. She draws from classical and neoclassical discussions of opium in order to contextualise divergent understandings of the drug by the late eighteenth and early nineteenth centuries when John Keats wrote his well-known poetry under the influence. By contrast to understandings of opium's drug effects that gained popularity in the early nineteenth century, such as those of John Brown or Thomas De Quincey, many writing during the period (including and especially Keats) still relied on classical and neoclassical understandings of the drug. This chapter explores ways in which drug-effect imagery in long-eighteenth-century medical books and poetry overlapped, presenting opium as inhibiting activity in both body and mind, specifically in the form of indolence, drowsiness, dulling and forgetfulness.

Irmtraud Huber's "'Grief's comforter, Joy's guardian, good King Poppy!': Opium and Victorian Poetry" considers questions of literary form alongside references to debates over the known drug effects of opium. She references debates stemming from contemporary medical texts, which disagreed about whether opium should be considered what they called a 'stimulant' (as in the case of most readings of Brown's system) or a narcotic. Huber argues that Victorian poetry paints a picture of attitudes towards the drug that differ from those that have emerged from many critical discussions of Victorian narrative fiction. She suggests that there might be a link between the cultural shift that increasingly privileged narrative fiction over poetry during the nineteenth century on the one hand and the rise of a discourse of addiction on the other. Drawing on authors such as Alfred Tennyson as well as less familiar figures, she locates this link in the way in which poetry and narrative fiction—and the novel in particular—invest to a different degree in the idea of the autonomous, rational individual who is in control of her or his actions and desires. Such a perspective highlights a genre politics that can be seen to lie at the heart of different literary attitudes towards the drug. Narrative fiction and lyric poetry thus may be said to show affinities with different aspects or interpretations of the drug's known effects.

In "Dangerous Literary Substances: Discourses of Drugs and Dependence in the Nineteenth-Century Sensation Novel Debates", Sarah Frühwirth shows that understanding drugs and their effects in the nineteenth century went hand-in-hand with a heightened awareness of their potential dangers and habit-forming qualities, leading also to widespread public concern. Nineteenth-century public anxieties not only centred on psychoactive substances—in particular, opium and alcohol—but also on other kinds of consumables: books. Owing to their sensational content and their alleged unhealthy effect on readers, sensation novels came under fire from all sides. In order to discredit the genre, reviewers frequently compared sensation fiction to opium and alcohol in order to refigure their readers as addicts. Similar to anti-drug campaigners, critics of sensation fiction focused on the pharmacodynamics of drugs in order to illustrate the dangers they posed to the reader's health and moral integrity, noting symptoms such as sweating, an increased heart and breathing rate and mental excitement as well as pointing to possible long-term adverse effects. Frühwirth further argues that, similar to the concerns voiced by anti-drug or anti-alcohol campaigners, reviewers' fears regarding 'drug effects' of this particular literary genre were not limited to individual bodies but also extended to the social and national body. In this light, the reviewers' recourse to the language of drug effect and addiction in their criticism of sensation fiction can be seen as an elaborate discursive strategy that enabled them to voice a wide variety of anxieties concerning social and national stability.

Rethinking the Pharmacological Body: Drugs and the Borders of the Human

In "Blurring Plant and Human Boundaries: Erasmus Darwin's *The Loves of the Plants*", C. A. Vaughn Cross explores Erasmus Darwin's role in the development of Western psychopharmacological knowledge. His first major literary work *The Loves of the Plants*, a poem which discussed drug experimentation, was published in 1789 and revised in 1792 as a literary jigsaw puzzle for non-specialists. This essay considers how Darwin's amalgamation of old and new paradigms of knowledge about plants, minerals and animal bodies sought to educate the public about potential pleasures, dangers and overall value of psychoactive substances, arguing that right use of various plants could contribute to the civic good, exploring the way drugs facilitate human and plant-interconnection alike. A popular health paradigm among Darwin's network of Edinburgh-trained physicians was Brunonianism, which looked for over- or under-stimulated, blocked or depleted nervous energy. One of several significant attempts of the era to advance knowledge of drug efficacy, Darwin's work stood out from contemporaries' efforts by combining Linnaeus' classificatory system with French naturalism, sensationalism, hermeticism and his own Baconian experiments and theories of psychophysiology. From recipes rendering single psychoactive entities edible to suggestive groupings of complex admixtures of liqueurs, gasses and powders containing psilocin, opiates or eugenol, *The Loves of the Plants* contains Darwin's prescriptions for health and happiness by interacting with psychoactive—'affecting'—substances, including warnings about doses and lethality. While the essay addresses Darwin's better-known contributions to opium use, it also takes into account his view of more than ninety potential drugs and new drug technologies such as pipes, chimneys and syringes for ingesting drugs such as cannabis, mushrooms and various forms of alcohol.

While figures such as John Brown, Thomas Beddoes and Thomas De Quincey had a lasting impact on how the public came to understand opium in particular, from today's perspective, such accounts of how opium works on the brain and the body are technically incorrect. In "Pharmacokinetics and Opium-Eating: Metabolites, Stomach Aches and the Afterlife of De Quincey's Addiction", Hannah Markley relies on current psychopharmacological understandings of how drugs act and, in turn, are acted on by the body to produce drug effects, drawing attention to the entanglement of action and effect (and of drug and human body) in modern scientific discussions of drugs. By focusing on the ways opiates are metabolised by the body, Markley argues that, despite his technically flawed accounts, De Quincey's rhetoric of 'opium-eating' in *Confessions* (1821) ironically anticipates current psychopharmacological explanations of how bodies and drugs interact to produce what is understood as a drug effect. Hence, recent psychopharmacological descriptions of drugs and their metabolisation help to confirm De Quincey's descriptions of the bilateral relationship between the opium user, or 'eater', and the drug itself. In this context, De Quincey's physical dependence is, in fact, the result of a complex entanglement of the mind, the body and the drug as they interact with prior maladies, eating habits and the consumption of other substances such as alcohol. Markley emphasises the interaction between opium and the stomach in particular by pointing out that opioid receptors are not only located in the brain but also in the gut. Most importantly, she adds that opiates must first be absorbed into the body before even having any recognisable drug effect on the central nervous system. What is more, the products of the body's digestive processes, *metabolites*, are no less potent than the drug itself. In this way, Markley discusses *pharmacokinetics*—the movements of drugs in a physiological system—to investigate the complex interactions that subtend the two most common pharmacological heuristics of drug action and drug effect. She then identifies how De Quincey's gastro-intestinal rhetoric remaps the interrelation of drug actions and effects as well as their side effects. Specifically, Markley traces how, in *Confessions*, the narrator's stomach pains are conflated with memories of starvation, emotional pain and drug withdrawal. In a pharmakonian sense, then, the drug becomes pain and relief, poison and remedy in ways that extend the pharmacological significance of 'opium-eating' to include the affective and psychological circuits in which De Quincey inscribes his habit. Markley concludes that while De Quincey's pharmacological explanations of addiction are inaccurate, they nonetheless hold possibilities for rethinking the social, cultural and biomedical contexts of our modern concepts of addiction.

One way of considering the relationship between humans and drugs is to rethink the way we imagine the boundary between the two. In "A Posthumanist Approach to Agency in De Quincey's *Confessions*", Anna

Rowntree takes a vital materialist approach arguing that substances, and not only humans, have a kind of agency. She points to De Quincey's own claim that opium is the 'true hero' of *Confessions*, taking this as her cue to de-centre the human and put pressure on the belief that the human is rational and self-governing. Rowntree's reconsideration of addiction takes into account the substance's power to affect the human body, arguing that the human and the drug act in assemblage with each other. Rowntree reads *Confessions* as a document of symptoms interesting to the medical practitioner, which points to the drug's power to exceed human comprehension and perhaps even empirical understanding.

Drug Action and Effect: The Cultural Politics of Psychopharmacological Knowledge

Our third section explores the way literary texts entice us to think about the way external forces—cultural and political—inflect the way contemporaries interpreted drug effects. In "Reading De Quinceyean Rhetoric Against the Grain: An Actor-Network-Theory Approach", Anuj Gupta considers the relationship between De Quincey's *Confessions* and nineteenth-century medical texts on opium by writers such as John Awister, John Jones and John Brown. He argues that the concept of 'anthropocentric utilitarianism' was the dominant discursive trope across medical and literary genres, a trope that conditioned the ways in which the nineteenth-century English public understood the drug effects of psychoactive substances. In other words, what mattered most in these texts was the question of which advantages or disadvantages drugs—in this case opium—had for the human body, whether they were useful and beneficial or whether they caused harm. In a rapidly industrialising age, these texts became increasingly occupied with the calming effects of opium, which led to the conceptualisation of the drug as an antidote for a culture in need of relief from the 'side effects' of capitalism. Gupta points out that this utilitarian orientation in De Quincey's writings and those of contemporary medical experts was limited by an anthropocentrism that imagined the interaction between humans and drugs as a one-way linear process involving an agential human being as subject and a lifeless, passive psychoactive substance as an object. He further argues that this rigid 'anthropocentric utilitarianism' is concomitant with a significant epistemic limitation of the modern Western world. In this regard, he calls attention to a so-called plant blindness characteristic to modern society, one that emerges in the nineteenth century. Gupta proposes that Actor-Network-Theory (ANT) offers a possible method to read these texts on opium against the grain in order to overcome these limitations. Using this method, he debunks the idea of opium as an inert substance and contextualises it with a complex history of human-plant co-evolution.

Another way economics informs psychopharmacology is through the way political economy and the Opium Wars are intertwined. In "Blood Streams, Cash Flows and Circulations of Desire: Psychopharmacological Knowledge About Opium in Nineteenth-Century Women's Fiction", Nadine Böhm-Schnitker focuses on the discourse on opium at the dawn of the First Opium War. Early nineteenth-century women's fiction about the domestic use of opium—whose effects are here understood within a Brunonian medical framework—cannot be read without considering global reverberations. Opium circulates in bloodstreams as well as in economic channels, and the logic of these circulations intersects with the social construction of gender and class inequalities at home. Nineteenth-century women writers document these intersections and betray the sociopolitical workings of opium by showing what kinds of psychological and physical relief the drug provides. Maria Edgeworth's *Belinda* (1801), for instance, foreshadows the concerns of later sensation fiction writers and documents the ways in which the psychological impact of opium betrays the intertwinements of the economic desires of empire, the cravings and addictions of the body, the structures of gendered suppression and political economy. Elizabeth Gaskell's *Mary Barton* (1853) connects the political with the private and reveals opium addiction to be a crucial relay between individual desire, empire and domesticity. The works of fiction under consideration here not only focus on the gendering of different forms of opium consumption but also render it abundantly clear that opium correlates with economic as well as colonial aspects that consequently interconnect the management of the body with the management of finances and colonial expansion. These texts provide a double reflection on psychopharmacology in that they represent the characters' knowledge about opium's effects on mind and body on the diegetic level and reveal the wider sociocultural contexts in which the drug plays a role.

Besides economics, national discourses also inflect the way contemporaries understood drugs. This is particularly telling in the case of absinthe, a drink made from alcohol and wormwood (whose active ingredient is a drug called thujone). In "The Indeterminate Pharmacology of Absinthe in Nineteenth-Century Literature and Beyond", Vanessa

Herrmann considers the known drug effects of absinthe in the Victorian period alongside today's scientific knowledge. She draws on results provided by twenty-first-century medico-scientific studies that are, interestingly, not consistent in their findings about thujone's drug effects. Some studies deny any hallucinogenic or lethal attributes of thujone while others confirm it. In order to highlight the drink's indeterminate effects, Herrmann provides a reading of two nineteenth-century literary texts—by Christina Rossetti and Robert Hichens—representing a common understanding of the beverage. Absinthe was the drink of choice for artists such as Wilde, Toulouse-Lautrec, Picasso and Van Gogh. Attitudes and interpretations of the drink's effects are sometimes ambiguous, and they reflect national attitudes (in this case, ones about the French) as well as moral outlooks.

Historicising the Prescription: Medication and Self-Medication

With the growing professionalisation of medicine came a more structured and systematic way of recommending and dispensing drugs, and a tighter definition of how drugs should be used as medications, but this was a long time coming. Our first two contributions in this section argue that gender plays a significant role in the way drugs were prescribed. In "'She furnishes the fan and the lavender water': Nervous Distress, Female Healers and Jane Austen's Herbal Medicine", Rebecca Spear draws on the question of gender to understand the way drugs were prescribed by laypeople. Although medicinal botany was becoming increasingly associated with professional medical practice, the dispensation of simple medicines for acute nervous distress remained synonymous with the medical care intrinsic to women's social interactions and private experiences. Spear states that a reference to lavender water—a common plant-based remedy for nervous distress in the eighteenth and nineteenth centuries—was deleted in the history of Austen's novella "Kitty, or the Bower" (1792–1816). In addition to power shifts from medicinal to scientific botany that affected women's medical practices and proto-feminist communities, Spear examines the effects of fashion and sexual difference upon the uses of lavender as a substance associated with strengthening the nerves. Advancing its discussion of nervous disease, sexual difference and the fashionable remedy, this contribution also takes Austen's Steventon novels *Northanger Abbey* (1818) and *Sense and Sensibility* (1811) into consideration. Austen's depictions of lavender's use, its soothing and restorative drug effects, function as a means of subverting hypotheses regarding women's predisposition to nervous disease. Discussing the experimental *materia medica*, Spear suggests that portrayals of lavender's potency serve as Austen's protest against patriarchal suppression through professionalisation of the medical practice and medicalisation of women's proto-pharmacological knowledge. She asserts that lavender—a substance regarded as efficacious in folk and rational medicines—becomes a metonym for the recovery of women healers, their knowledge and proto-feminist communities in Austen's writings.

Joseph Crawford's "'When poor mama long restless lies, / She drinks the poppy's juice': Opium and Gender in British Romantic Literature" brings the question of domesticity to bear on prescription and gender. By the later eighteenth century, it had become clear that opium could function either as a 'stimulant' or as a 'relaxant'. These different drug effects, however, did not have equal cultural status, and while the use of opium as a sedative and painkiller was routine and unremarkable, the recreational use of the drug for its stimulant and hallucinogenic properties was the subject of widespread disapproval and was associated with the deviant figure of the 'oriental opium-eater'. Samuel Taylor Coleridge and Thomas De Quincey both wrote about the vision-inducing powers of opium, and its importance to their respective works has been repeatedly investigated by scholars. It has more seldom been noted, however, that another literature of opium use existed in Romantic-era Britain, which is to be found in the less well-known works of their female contemporaries. Opium literature written by women depicted the drug in a very different light: not as a seductive and destructive stimulant which enticed the (male) artist away from his domestic responsibilities, but as a relaxant that helped its (female) users to cope with the demands imposed by the very world of mundane domesticity which male opium-eaters were stereotypically regarded as using the drug to forsake. Crawford suggests that this distinction between male and female writing about opium in the period corresponds to the double role played by the drug in contemporary British society as a whole. By aligning their use of opium with the 'moderate' and medicinal consumption of opiates recommended by contemporary doctors rather than with the deviant figure of the selfish and hedonistic opium-eater, these female writers were able to imply that their drug use was compatible with, and indeed contributed to, their ability to selflessly and effectively discharge their domestic duties.

Our final essay marks a transition from self-medication and prescription to the rise of a more regulated professional medicine and pharmacy. Björn Bosserhoff's "*Middlemarch* and Medical Practice in the Regency Era: From 'Bottles of Stuff' to the Clinical Gaze" argues that during much of the nineteenth century, rather than being perceived as an exotic recreational drug, opium was very much part of the daily lives of Britons—usually taken in the form of laudanum, a wildly popular tincture prescribed for anything from common coughs and colds to tuberculosis and 'insanity'. In his discussion of George Eliot's *Middlemarch*, Bosserhoff focuses on the novel's doctor-protagonist, Tertius Lydgate, and his stance towards the use of drugs in medical practice. *Middlemarch* portrays a time when, though little was actually known about drug action, most British practitioners nonetheless resorted to shotgun polypharmacy as treatment of choice for most diseases. Trained in Paris, Lydgate instead champions a largely noninterventionist approach, based—as Bosserhoff shows by contextualising Lydgate's views within contemporary discourses on pharmacotherapy—on models such as Vitet's and Laënnec's *médecine expectante*. Eliot's fictional doctor thus emerges as one of the foremost examples of an early clinician in the Foucauldian mode, a pioneering scientist-physician with a greater understanding of the effects drugs like opium have on individual bodies.

In 2017, the philosopher Fredric Svenaeus published a study on phenomenology and bioethics in which he argued that medical technologies, among them psychopharmacological drugs, have altered our understanding of what it means to be human. Svenaeus calls for a philosophical analysis of this question and criticises the absence of such studies in the field of medical ethics. In the same way that Svenaeus' book brings phenomenology and bioethics together, our volume interweaves psychopharmacology and literature in order to tackle questions that often get left out of public discussion of drugs owing to disciplinary boundaries. Literary and cultural texts help to do the work that Svenaeus and others have been calling for: to consider the social, cultural and political contexts alongside the science of drugs, a context which forces one to address the nuances of interpreting how one feels under the influence, and how the drug effect is situated within a nexus of other forces. Such work is tantamount to parsing out the human condition and is therefore a relevant supplement for other cultural practices fused with the allure and promises of technological progress. It is this sort of insight that the present volume hopes to create by focusing on literature from the last decades of the eighteenth century to the close of the nineteenth, from Erasmus Darwin to the use of absinthe in the *fin de siècle*.

References

Abrams, M. H. 1971. *The Milk of Paradise: The Effect of Opium Visions on the Works of De Quincey, Crabbe, Francis Thompson, and Coleridge*. New York: Octagon Books.

Barbara, Jean-Gaël. 2015. History of Psychopharmacology: From Functional Restitution to Functional Enhancement. In *Handbook of Neuroethics*, 1st ed., ed. Jens Clausen and Neil Levy, 489–504. Dordrecht: Springer Science + Business Media.

Boon, Marcus. 2002. *The Road of Excess: A History of Writers on Drugs*. Cambridge: Harvard University Press.

Budge, Gavin. 2013. *Romanticism, Medicine and the Natural Supernatural: Transcendent Vision and Bodily Spectres, 1789–1852*. Basingstoke: Palgrave Macmillan.

Campbell, Elizabeth A. 2007. Don't Say It with Nightshade: Sentimental Botany and the Natural History of 'Atropa Belladonna'. *Victorian Literature and Culture* 35 (2): 607–15.

Clej, Alina. 1995. *A Genealogy of the Modern Self: Thomas De Quincey and the Intoxication of Writing*. Stanford: Stanford University Press.

Comitini, Patricia. 2012. The Strange Case of Addiction in Robert Louis Stevenson's *Strange Case of Dr. Jekyll and Mr. Hyde*. *Victorian Review* 38 (1): 113–31.

Courtwright, David T. 2001. *Forces of Habit: Drugs and the Making of the Modern World*. Cambridge: Harvard University Press.

Curth, Louise Hill. 2006. *From Physick to Pharmacology: Five Hundred Years of British Drug Retailing*. Aldershot: Ashgate.

Davenport-Hines, Richard. 2001. *The Pursuit of Oblivion: A Social History of Drugs*. London: Weidenfeld & Nicolson History.

Davy, Humphrey. 1858. *Fragmentary Remains, Literary and Scientific*, ed. John Davy. London: John Churchill.

De Boever, Arne. 2013. *Narrative Care: Biopolitics and the Novel*. New York: Bloomsbury.

Derrida, Jacques. 1981. Plato's Pharmacy. In *Dissemination*, trans. Barbara Johnson 429–50. London: University of Chicago Press.

Drews, Jürgen. 2000. Drug Discovery: A Historical Perspective. *Science* 17: 1960–1964.

Foxcroft, Louise. 2016. *The Making of Addiction: The "Use and Abuse" of Opium in Nineteenth-Century Britain*. Abingdon: Routledge.

Francia, Susan, and Anne Stobart. 2015. *Critical Approaches to the History of Western Herbal Medicine*. London: Bloomsbury.

Franke, Andreas Günter, et al. 2015. The Case of Pharmacological Neuroenhancement: Medical, Judicial and Ethical Aspects from a German Perspective. *Pharmacopsychiatry* 48 (7): 256–64.

Herlinghaus, Hermann. 2018. *The Pharmakon: Concept Figure, Image of Transgression, Poetic Practice*. Heidelberg: Universitätsverlag Winter.

Jay, Mike. 2011. *High Society – Eine Kulturgeschichte der Drogen*. Darmstadt: Primus.

Jenkins, Janis H. 2011. *Pharmaceutical Self: The Global Shaping of Experience in an Age of Psychopharmacology*. Santa Fe: School for Advanced Research Press.

Maier, Larissa, and Michel P. Schaub. 2015. The Use of Prescription Drugs and Drugs of Abuse for Neuroenhancement in Europe: Not Widespread But a Reality. *European Psychologist* 20 (3:) 155–66.

Malleck, Daniel. 2020. *Drugs, Alcohol and Addiction in the Long Nineteenth Century*. 4 Vols. Milton Park: Routledge.

Mangiavellano, Daniel R. 2013. De Quincey, Coleridge, and the Literary Model of Habit. *Nineteenth-Century Prose* 40 (1): 27–60.

Martin, Alison E. 2011. The Voice of Nature: British Women Translating Botany in the Early Nineteenth Century. In *Translating Women*, 1st ed., ed. Luise von Flotow, 11–35. Ottawa: University of Ottawa Press.

Milligan, Barry. 1995. *Pleasures and Pains: Opium and the Orient in Nineteenth-Century British Culture*. Charlottesville: University Press of Virginia.

———. 2005. Morphine-Addicted Doctors, the English Opium-Eater, and Embattled Medical Authority. *Victorian Literature and Culture* 33 (2): 541–53.

Morrison, Robert. 2011. De Quincey's Addiction. *Romanticism: The Journal of Romantic Culture and Criticism* 17 (3): 270–77.

Morrison, Robert, and Daniel Sanjiv Roberts. 2008. *Thomas De Quincey: New Theoretical and Critical Directions*. New York: Routledge.

Partridge, Christopher. 2018. *High Culture: Drugs, Mysticism, and the Pursuit of Transcendence in the Modern World*. New York: Oxford University Press.

Petryna, Adriana, et al. 2006. *Global Pharmaceuticals—Ethics, Markets, Practices*. Durham: Duke University Press.

Pies, Ronald. 2006. The Complexities of Psychopharmacology. Psychiatric Times. *psychiatrictimes*.com. Accessed 8 August 2018.

Porter, Roy. 1988. Brunonian Psychiatry. In *Brunonianism in Britain and Europe*, ed. W.F. Bynum and Roy Porter, 89–99. London: Wellcome Institute for the History of Medicine.
Porter, Roy, and Mikulas Teich. *Drugs and Narcotics in History*. Cambridge: Cambridge University Press, 1995.
Racine, Eric, and Cynthia Forlini. 2010. Cognitive Enhancement, Lifestyle Choice or Misuse of Prescription Drugs? Ethic Blind Spots in Current Debates. *Neuroethics* 3 (1): 1–4.
Redfield, Marc, and Janet Brodie. 2002. *High Anxieties—Cultural Studies in Addiction*. Berkley, Los Angeles, and London: University of California Press.
Reed, Thomas. 2006. *The Transforming Draught: Jekyll and Hyde, Robert Louis Stevenson and the Victorian Alcohol Debate*. Jefferson: McFarland.
Ronell, Avital. 2004. *Crack Wars: Literature, Addiction, Mania*. Urbana: University of Illinois.
Ruston, Sharon. 2012. *Shelley and Vitality*. Basingstoke: Palgrave Macmillan.
Rzepka, Charles J. 1991. The Literature of Power and the Imperial Will: De Quincey's Opium War Essays. *South Central Review* 8 (1): 37–45.
Schiebinger, Londa. 2005. Prospecting for Drugs: European Naturalists in the West Indies. In *Colonial Botany: Science, Commerce, and Politics in the Early Modern World*, ed. Londa Schiebinger and Claudia Swan, 119–33. Philadelphia: University of Pennsylvania Press.
———. 2007. *Plants and Empire—Colonial Bioprospecting in the Atlantic World*. Cambridge: Harvard University Press.
———. 2017. *Secret Cures of Slaves: People, Plants, and Medicine in the Eighteenth-Century Atlantic World*. Stanford: Stanford University Press.
Schiller, Francis. 1976. Thomas de Quincey's Lifelong Addiction. *Perspectives in Biology and Medicine* 20 (1): 131–41.
Servitje, Lorenzo. 2018. Of Drugs and Droogs: Cultural Dynamics, Psychopharmacology, and Neuroscience in Anthony Burgess's *A Clockwork Orange*. *Literature and Medicine* 36 (1): 101–23.
Small, Douglas. 2016. Masters of Healing: Cocaine and the Ideal of the Victorian Medical Man. *Journal of Victorian Culture* 21 (1): 3–20.
Stiegler, Bernard. 2011. Pharmacology of Spirit and That Which Makes Life Worth Living. In *Theory After Theory*, ed. Jane Elliott and Derek Attridge. Abingdon: Routledge.
Stiles, Anne. 2007. Introduction. In *Neurology and Literature, 1860–1920*, ed. Anne Stiles, 1–23. Basingstoke: Palgrave.
Svenaeus, Fredrik. 2017. *Phenomenological Bioethics Medical Technologies, Human Suffering, and the Meaning of Being Alive*. Abingdon: Routledge.
Vice, Sue, et al. 1994. *Beyond the Pleasure Dome: Writing and Addiction from the Romantics*. Sheffield: Sheffield Academic Press.

Walezak, Emilie. 2018. A. S. Byatt, Science, and the Mind/Body Dilemma. *Journal of Literature and Science* 11 (1): 106–19.
Wallis, P. 2012. Exotic Drugs and English Medicine: England's Drug Trade, c. 1550–c. 1800. *Social History of Medicine* 25 (1): 20–46.
Weber, Leonard J. 2006. *Profits Before People? Ethical Standards and the Marketing of Prescription Drugs*. Bloomington: Indiana University Press.
Zieger, Susan. 2008. *Inventing the Addict: Drugs, Race, and Sexuality in Nineteenth-Century British and American Literature*. Amherst: University of Massachusetts Press.

PART I

Drugs and Genre

CHAPTER 2

Historicising Keats' Opium Imagery Through Neoclassical Medical and Literary Discourses

Octavia Cox

On 18 March 1819, John Keats was playing cricket when "disaster" struck, the ball conked him "dir[ectl]y on the sight" giving him "a black eye" (Keats 1958, 2: 78). In addition to applying "a lee[ch to] the eyelid to reduce the "infla[mm]ation", in all probability, given its medical status as the most efficacious pain relief, Keats dosed himself with opium; his "indolent" mood the next morning certainly suggests that he was effected by the drug (Keats 1958, 2: 78; see Ober 1968, 872; Roe 2012, 308). Being the foremost poet-physician of the period, it is unsurprising that Keats' professional knowledge filtered into his poetical imagery.[1] As this essay shows, Keats' use of opium imagery was in line with neoclassical medical and literary understandings of opium. Thomas

[1]Some critics, however, have overlooked Keats' use of opium imagery in his poetry: for example, that Keats' "writing does not indicate any particular interest in the drug" (Boon 2002, 31).

O. Cox (✉)
University of Nottingham, Nottingham, UK
e-mail: Octavia.Cox1@nottingham.ac.uk

N. Roxburgh and J. S. Henke (eds.), *Psychopharmacology in British Literature and Culture, 1780–1900*, Palgrave Studies in Literature, Science and Medicine,
https://doi.org/10.1007/978-3-030-53598-8_2

De Quincey, on the other hand, whose *Confessions of an English Opium-Eater* (1821) rendered him, practically overnight, a de facto opium expert in the eyes of the nation, felt frustrated at previous medical and literary accounts, denouncing "all that has been hitherto written on the subject of opium" as "Lies! lies! lies!" (De Quincey 1998, 39). Some have taken De Quincey's account to be representative of historical attitudes towards opium, which obscures the neoclassical tradition inherited by Keats. This essay, therefore, attempts to recover what neoclassical medical and literary discourses conveyed about opium's effects in the light of the way they later became obscured through the prevalence of De Quincey's understanding in the 1820s and beyond. In so doing, it examines the interplay between poetic imagery alongside emergent therapeutic description, showing that imagery and diction about opium remained consistent across genres, both literary and medical.

As Andreas-Holger Maehle notes, there was "small psychopharmacological interest" in opium's effects before the nineteenth century (Maehle 1999, 188; see Davenport-Hines 2004, 54). Eighteenth-century physicians, and other scientific experimenters, were principally interested in discovering, and understanding, opium's pharmacological effects—so, for instance, whether opium operated on the nervous or cardiovascular system.[2] Any psychopharmacological or psychological effects that were noted were usually mentioned only incidentally (see Jones 1700, 21–24; Leigh 1786, 113–14; Crumpe 1793, 45–46). The lack of a methodological basis for result-gathering hindered psychopharmacological inquiry (Maehle 1995, 68). Vivisectional physiology had long provided a basis for somatic experimentation with opium—Samuel Pepys, in his diary for 16 May 1664, for instance, reports attending an experiment in which a dog was injected with opium inducing "a strange and sudden effect" (5: 151); while Robert Whytt, President of the Royal College of Physicians, in the mid-eighteenth century, details his "Experiments made with Opium on Living and Dying Animals" (1768, 309–27)—but there were yet not yet specific methods known to examine opium's psychological

[2] For example: "*It is observed by all, that* [opium] *mainly affects the* Genus Nervosum, *and animal Spirits*, and not the *Bloud* and *Humors*" (Jones 1700, 24) and "Effects will all be heightened by the Mixture of the *Opiate* Particles with the *Blood*; Which is hereupon Rarefied, and Distends its Vessels, especially those of the Brain" (Mead 1702, 143).

effects beyond self-experimentation and anecdotes. Literary writers, therefore, were in the same position as medical experimenters when it came to understanding opium's psychopharmacological effects.

This essay argues that classical conceptions of opium filtered into, almost without contestation, neoclassical perceptions, both in medicine and in literature. It historicises opium's qualities as a widespread pain-relieving drug in long-eighteenth-century culture, taken to soothe both physical and mental ills, and demonstrates that many writers depicted their everyday experiences of opium's effects in keeping with medical observations. Indeed, opium imagery aligns in a multitude of genres: from scientific medical disquisition, to poetry, to fiction. The chapter shows that Keats' imagery accords with earlier accounts of opium's psychopharmacological effects, especially in conceiving of opium as bodily and mental pain relief used primarily to procure sleep and drowsiness, to promote a dulling and lulling of sensation, to provide a pleasant oblivion and forgetfulness, and as a soothing balm; while acknowledging that it induces indolence, stupor and insensibility. These were terms used across medical and literary discourses. Thus Keats, along with a preponderance of post-1780 but pre-De Quinceyan writers, rejected 1780s Brunonian theory, which stood in opposition to the neoclassical account. Brunonian theory articulated that opium's "great" "virtue" is the "higher degree" of its "stimulant powers", that although there are some "circumstances in which opium produces sleep", primarily "it excites the functions of body and mind" (Brown 1795, 1: 282–85). De Quincey's Brunonian sympathies (Milligan 2007) are evident in his conviction that "the primary effects of opium are always, and in the highest degree, to excite and stimulate the system" (De Quincey 1998, 43–44). During lectures at Guy's hospital, Keats would have heard of the "Speciousness", "embarrassments" and "farther difficulties of [Brown's] doctrine" (Babington and Curry 1811, 21–22).

Classical Precedents for Long-Eighteenth-Century Conceptions of Opium

Homer's poetry contained the first reference in Greek literature to the opium poppy (Scarborough 1995, 4), an image which subsequently appeared in Virgil's *Aeneid*. Alexander Pope, in the *Odyssey* (1725–1726), translated Homer's image thus: "Bright *Helen* mix'd a mirth-inspiring bowl [...] / Charm'd with that virtuous draught, th' exalted mind / All sense of woe delivers to the wind" (Bk.IV, ll.302–08). "The conjectures about this cordial of *Helen* have been almost infinite", Pope commented in his footnote to the lines above, but "[t]he description of this *Nepenthes* agrees admirably with what we know of the qualities and effects of *Opium*" (9: 133 footnote). Centuries before the term *psychopharmacology* was coined, then, Pope described the effects of Helen's opium-poppy "drug" on "th' exalted mind"; it renders its imbibers "impassive and serene" as a "man entranc'd" (Bk.IV, ll.303–14). Helen uses the drug to make those at Menelaus' feast forget the pain, "All sense of woe", caused by the Trojan War. Millennia before medical and scientific experimentation on opium's psychopharmacological processes began, then, Homer had already described its effects on the mind: "a drug to quiet all pain and strife, and bring forgetfulness of every ill" (Murray 2014, 1: 135).

By the eighteenth century, opium was renowned for sending its imbibers to sleep; indeed, the poppy species' scientific classification, *Papaver Somniferum*—coined by Carl Linnaeus in the mid-eighteenth century, in *Species Plantarum* (1753)—translates from Latin as *sleep-inducing poppy*. Its soporific qualities are, therefore, embedded within its very scientific identity. Linnaeus echoed Virgil, who documented the "soporiferumque papaver" in the *Aeneid*, when Dido recalls "a priestess" who performs "spells" and "magic arts" by "sprinkling [...] slumbrous poppies" (Fairclough 2014, 1: 454–55). Virgil's image filtered into English vernacular through the neoclassical era's translations: in 1697, into "Poppy-Seeds" that "sooth'd" "into sleep" (Dryden 1956–1989, 5: 475); into "drowsy Poppy's Juice" (Trapp 1731, 2: 260); and into "poppy's soporiferous grain" (Beresford 1794, 144).

In accordance with classical precedent, four English canon stalwarts—Geoffrey Chaucer, Edmund Spenser, William Shakespeare and John Milton—all used opium imagery in association with soporifics, drowsiness and mental soothing. As early as the fourteenth century, Chaucer, in the "Legend of Good Women", used the image of "narcotyks and opies" to procure "slepe as longe as evere thee leste ['please']" (ll.2669–70), and in the "Knight's Tale", Palamon drugs his jailer with "nercotikes and opie", so "That all that nyght, thogh that men wolde him shake, |

The gayler sleep; he myghte nat awake" (ll.1472–74).[3] Moving forward to the sixteenth and seventeenth centuries, in the *Faerie Queene* (1590), Spenser wrote of the "Dead sleeping *Poppy*" by which "Wise *Socrates*" was "made to dy" (II vii st.52): as though opium elicits oblivion, even in the greatest of minds. In Shakespeare's *Othello* (c.1603), Iago predicts that the "poison" he drips into Othello's ear will sufficiently agitate Othello that even the most powerful opium would produce no rest:

> Not poppy, nor mandragora,
> Nor all the drowsy syrups of the world
> Shall ever medicine thee to that sweet sleep
> Which thou owed'st yesterday. (III iii ll.361–69)

Iago's words are "poison" to Othello's thoughts, against which even the "drowsy syrup" of "poppy" cannot "medicine" Othello. Meanwhile, in "Samson Agonistes" (1671), Milton described "death's benumbing opium" as the "only cure" for the "torment" of "the inmost mind", which no other "med'cinal liquor can assuage" (ll.606–30). As we progress through these four writers, the way opium is perceived changes. In Chaucer and Spenser, it is a poison with whose aid to harm others; by contrast, in seventeenth-century texts, opium is associated with therapeutic analgesics and assuaging mental torment. Crucially, seventeenth-century texts connect opium expressly with "medicine" and the "med'cinal", referencing the period's developing medical community (Ackerknecht 2016). Even as early as 1640, John Parkinson[4] defined opium as "*Narcoticke* medicine" (368).

[3] In both instances, Chaucer makes reference to opium explicitly, which his source material had not: Ovid's *Heroides* 14 (Emerson 1919, 287) and Giovanni Boccaccio's *Teseida delle nozze d'Emilia* (Chaucer 2008, 826). Ovid merely referred to wine's soporific effects: "Yourself lay quiet in the grip of the sleep the wine had given you" (172–73). In *Teseida* (Book V), Palamon daringly escapes from prison by changing clothes with Alimeto, his physician. In Chaucer's revision, instead of a physician assisting Palamon's escape, opium enables the absconsion.

[4] Founding member of the Worshipful Society of Apothecaries (1617), apothecary to James I and Royal Botanist to Charles I.

Historicising Opium as Medicine

Pope's Homeric description, of Helen's "drugs, so friendly to the joys of life" (Bk.IV, l.315), differs markedly from Thomas De Quincey's ultimately angst-inducing account, a century later, of "a mind debilitated by opium" (1998, 79). *Confessions* has since been considered the *locus classicus* of narcotic experience (Ober 1968, 862; Hayter 1968, 36), helping to contribute to Victorian anxieties about opium's threat to Britain (Schmitt 2002). Yet before the nineteenth century, opium consumption was considered unremarkable, even mundane; despite its associations with the East, opium was not then framed as a threatening Other (Milligan 1995; Leask 1992). Unlike Samuel Taylor Coleridge's *Kubla Khan* (1816), De Quincey's *Confessions*, and later texts which followed in the same vein, neoclassical writing tended to reflect the period's popular discourse, which staged opium as ordinary, common and domestic (Freeman 2012). Opium sat in the medicine cabinets of vast swathes of the population, was valued for its pain-alleviating and sleep-inducing qualities, and was prescribed for all manner of complaints (Holloway 1995, 82). Alexander Philip Wilson's observation, in *An Experimental Essay, on the Manner in which Opium Acts* (1795), that "most people have experienced [opium's effects], indeed, in their own persons" indicates how prevalent opium consumption was (3). Opium was medicine, rather than, as subsequently, opium use being thought a reason to seek medical help (Berridge and Edwards 1987, 13–14). It was not until 1826, five years after *Confessions* first appeared, that the first modern experimental psychopharmacological text was produced, Pierre-Alexandre Charvet's *De l'action comparée de l'opium* (Siegel and Hirschman 1983). Others have located the scientific beginnings of modern psychopharmacology later, in 1892 with the work of Emil Kraepelin (Maehle 1999, 188). Those writing of opium's psychological effects before then, from both a medical and literary perspective, had to rely for explanation on classical imagery, accounts from "*Eastern Nations*" (Jones 1700, 21), and/or self-experimentation, self-experience and anecdotal evidence.

Authors' choices of diction in translating classical texts conform to the seventeenth- and eighteenth-century view of opium as medicine. Seventeenth-century renderings of Homer's *Odyssey* referred to Helen's nepenthe as "usefull med[i]cine" (Chapman 1616, 52), "this Med'cine" used "to appease all woe" (Ogilby 1665, 45) and "*Helens* Medicine" (Hobbes 1675, 41). A century later, the drug was still presented mainly in terms of its medicinal properties: Helen "medicated" the wine (Cowper 1791, 79), and administered "medicating drops" (Sotheby 1834, 3: 83). In later translations—those written after the Opium Wars of 1839–1842 and 1856–1858, when public disquiet over opium use increased—explicitly medicinal terminology is omitted: they are deemed "cunning and excellent drugs" (Buckley 1851, 49); and the "anodyne" effect of Helen's "potent drugs" is noted (Alford 1861, 55). The Victorian-period tendency not to characterise Helen's opium-like drug as medicine continues in modern translations: Loeb terms them "cunning drugs" "of healing" (Murray 2014, 1: 137), while *Oxford World's Classics* uses "cunning powerful drugs" (Shewring 1998, 35–54).

In addition to translations, descriptions by herbalists and physicians also evidence the neoclassical medicinal designations of opium: Nicholas Culpeper warned of laudanum, "have a care how you be too busy with such medicines lest you make a man sleep till dooms-day" (1649, 240); John Pechey refers to "Opiate Medicines" (Sydenham 1696, 140); Richard Mead discusses "the chief Virtues of this Medicine" (1702, 132); Thomas Cooper deems opium "a Medicine of great Power and Efficacy" (1766, 277); Alexander Monro refers to "powerful Medicines, viz. Opium" (1771, 292); Robert Hamilton declared it had a "deserved estimation in Medicine" (1790, 4). Even in texts that did not expressly refer to opium as a medicine, medical terminology was still deployed. Robert Burton's influential *Anatomy of Melancholy* (1621), for instance, advises that one should "take" the "remedy" and "Physicke" of "opiats, syrup of Poppy" in order to "procure sleepe", which would alleviate the "symptome[s]" of "dry braines" (2: 255). Importantly, Burton advocates opium for the cure of, what we now consider, mental unhealth: "continuall cares, feares, sorrowes [...] much crucifies melancholy men, and must therefore bee speedily helped" (2: 225). Elsewhere, opium was seen as a curative for physical ailments, such as dysentery: in the 1660s, pioneering physician Thomas Sydenham had prescribed laudanum for the euphemistically named "*Bloody Flux*" (1696, 125–44).

Throughout the long eighteenth century, opium was described as being taken in "doses"—"a moderate dose of opium" (Mead 1702, 141; Hill 1751, 782; Wilson 1795, 3); "large doses of opium" (Kirkland 1780, 63); "an over-dose of Opium" (Crumpe 1793, 45)—reiterating the contemporary categorisation of it as medicine. Samuel Johnson, in his *Dictionary of the English Language* (1755), defined "dose" in explicitly medical terms, as "So much of any medicine as is taken at one time", and the entry for "opium" referred repeatedly to "a dose of *opium*". A lecture on opium at Guy's hospital, which was part of Keats' training, refers to the "Extent and frequency of the dose most proper in general" (Babington and Curry 1811, 43). Even those in literary rather than medical circles referred to opium in "doses". Thomas Brown wrote jauntily, in 1685, of opium's sense-numbing effects: "How such a fit of Lethargy, / My Senses has possest, / As if a Dose of Opium / Had buried me in rest!" (Brown 1715, ll.1–4). An article by 'O' in the *English Review* (June 1785) referred to the "profound temporary lethargy, such as patients experience from large doses of opium" (5: 454). Similarly, Sara Hutchinson used imagery of medical prescription when she reported, in November 1815, that De Quincey was "often tipsey" because "[h]e doses himself with Opium" (88). De Quincey, however, disparaged such professional language: "the dose (to speak medically)" (1998, 44).

Self-medicating with opium was standard practice (Holloway 1995, 82). Lady Mary Wortley Montagu, in *Spectator* no. 573 (28 July 1714), could write blithely of a man who "composed himself with a Dose of Opium" (74). After a long day's travelling in 1798, Jane Austen's mother took "12 drops of Laudanum when she went to Bed, as a Composer" (1995, 16). By the nineteenth century, opium was "commonly and unselfconsciously bought and used" in "all levels of society" (Berridge and Edwards 1987, 49). Indeed, widespread opium use was sanctioned by the government, which kept its import duty low at 9 shillings per pound until 1826 (Wu 2015, 244). Through the long eighteenth century, then, opium became widely classified and endorsed as medication used to treat a multitude of maladies, bodily and mental, owing to known drug effects (even though opium's recognised effects were not yet physiologically chemically proven).

Keats' Opium Imagery in "To Autumn"

Registered, in October 1815, as a Surgeon's Pupil at Guy's and St Thomas' hospitals, enrolling for lectures on "Chemistry, Practice of Medicine, Theory of Medicine and Materia Medica" (Ghosh and Roe 2017, 25), Keats received his certificate to practice medicine the subsequent year. Keats could not have avoided being cognizant of opium's effects. Indeed, a whole lecture, no. 155, on Guy's Practice of Medicine course, was dedicated to "OPIUM". Guy's physicians considered the "[a]bstract question of [opium] being *stimulant* or *sedative*, improper", because, in contrast to Brunonian excitability theory, "its power of allaying excessive action" is "incontestable": "Doctrines which inculcate its uniform operation as a stimulus, therefore, fallacious" (Babington and Curry 1811, 42).[5] Moreover, Keats had personal, as well as professional, knowledge of opium's effects. In September 1819, Keats was "secretly taking" opium "to keep up his spirits" (Brown 1937, 63), perhaps as a remedy for his developing tuberculosis (see Babington's and Curry's lecture no. 439, which considered opium as a possible treatment). Charles Armitage Brown further remarked ominously that "[Keats] needed not to be warned of the danger of such a habit" (63). In neoclassical medical theory, opium was not taken in order to excite stimulation (Brunonianism), but rather to achieve mental relief by balancing spirits' equilibrium: the supposed beneficial "*effects of Opium in lowness of spirits*" had long been noted—the physician George Young had observed in 1753 that giving a patient "his dose of opium [...] set his mind at ease" (101). Similarly, a character in Maria Edgeworth's novel *Belinda*[6] (1801) could be described as having "completely repaired [...] her spirits with opium" (34).

"How beautiful the season is now", wrote Keats to John Hamilton Reynolds, 21 September 1819—the very time that Brown had worried about Keats' opium consumption—explaining how "this struck [him] so much" he "composed upon it" (2: 167)—the poem became "To Autumn". The poem is infused with medical diction. In the original manuscript in his hand, Keats had initially written "Dosed with red[7] poppies", which he scribbled out, and replaced with "Dos'd with the fume of poppies" (Hebron 2009, 143). Charles Armitage Brown transcribed "Dosed" in his copy (Keats 1978, 669). Jack Stillinger suggests that Keats' and Brown's "Dos'd/Dosed" were "probably variant spellings of 'Dozed'" (Keats 1978, 670). Given Keats' medical training, being medicated with opium to sleep—from "Dose" to "Doze"—makes sense, punning on "dos/ze". In the same week as "To Autumn" was composed, Keats wrote to his brother George of his joy in "Pun-making" (Keats 1958, 2: 214). And two days after the poem's composition, Keats wrote

[5] Although published in 1811, *Outlines of Lectures* remained in use during Keats' time at Guy's Hospital, evident from the inscription inside the front cover in King's College London's copy, "Robert Pughe 1817".

[6] See essay in this volume by Joseph Crawford.

[7] See Stillinger (Keats 1978, 477) and Bate (2014, 183).

playfully to Richard Woodhouse, referring to his letter as a "dose" of himself, punning on Virgil's *Eclogues*, and transcribing the relevant line in "To Autumn" as "Dased with the fume of poppies" (Keats 1958, 2: 169–70). Keats later settled on "Drows'd" for the published 1820 version (l.17). In Keats' mind, then, a "Dose" of opium punningly moves between leaving one dozed, dazed and drowsed.

That Keats' first resort in "To Autumn" was to "Dosed", with its medical connotations, is consistent with the surrounding language, "asleep" and "slumberous": "sound asleep / Dosed with red poppies; while thy reeping hook / Spares form some slumberous / minutes while warm slumbers creep" (Hebron 2009, 143). Being "Dosed" with poppies led naturally in Keats' mind to being "Drows'd"; just as medically, one leads to the other. Autumn's "careless" mood here (l.14) echoes Keats' description of his own "supremely careless" post-opium "temper" the day after the cricket incident. Growing "careless" had long been considered an effect of opium-taking, noted, among others, by John Leigh (1786, 114).

In November 1820, a few months after Keats' volume was published, Byron associated Keats' poetry with "a Bedlam vision produced by raw pork and opium" (7: 226). But this is hardly apt to describe Keats' opium imagery here, which does not connote mad hallucination, but rather the opposite: Keats' poem is in keeping with opium's neoclassical medical associations of lulling and sedation. Althea Hayter deems Keats' use of opium imagery "mostly symbolic" (314), and his references to poppies unquestionably have literary symbolism. Nonetheless, they are also fully in line with neoclassical medical theory and practice. Indeed, the preoccupations we find repeated throughout Keats' poetry—indolence; being embalmed from pain; sleep and drowsiness; dulling and lulling; oblivion and forgetfulness—arguably have their roots in eighteenth-century medical culture surrounding opium.

Neoclassical Opium Imagery

Indolence, Stupor, and Insensibility

The etymological root of *indolence* is the state of non-pain, mental and physical—in essence, to be un-doleful. Describing his condition on 19 March 1819, Keats declared: "My passions are all asleep"; opium's chemical effects had "weakened [...] the fibres of the brain", "with the rest of the body", to make them "relaxed", and his "temper indolent and supremely careless" (Keats 1958, 2: 78). Jones found that opium seized one "not unlike the gentle, sweet, *Deliquium* that we find upon our entrance into a most agreeable *Slumber*" (1700, 20), while Keats' opium-induced "slumber" leaves "a delightful sensation about three degrees on this side of faintness" (Keats 1958, 2: 78).

Keats' language describing the episode is steeped in medical terminology (Goellnicht 1984, 205). Physician-poet Samuel Garth[8] had, in his poem *Dispensary* (1699), similarly described someone "[w]ith *Poppys* [...] o're him spread" as being in a "careless" state, with "His leaden Limbs at gentle ease," and aligned with "The God of Sloth" (4–5). Another physician-poet, Erasmus Darwin, produced *Loves of the Plants* (1789), which was "design[ed]" "to inlist the Imagination under the banner of Science" (ix), by describing remedies literarily in the poem and medically in the footnotes. Garth's poem had described the poppy-user who "supinely nods" "Upon a Couch of Down" in his/her "Abodes" (5); Darwin's echoes Garth's poem:

> Sopha'd on silk, amid her charm-built towers,
> Her meads of asphodel, and amaranth bowers,
> Where Sleep and Silence guard the soft abodes,
> In sullen apathy PAPAVER nods. (Darwin 1789, 69)

The decidedly poetic, neoclassic "asphodel" and "amaranth" suggest the mental ease and calm, the "charm[s]", procured by opium, while the drug simultaneously induces bodily apathetic inactivity. In the footnotes, Darwin observed, partially in line with Brunonian theory: "In small quantities [opium] exhilarates the mind, raises the passions, and invigorates the body", but he also warned, "in large ones it is succeeded by intoxication, languor, stupor and death". Darwin's footnotes use medical description while the poetry anthropomorphises Papaver, but both agree in their diagnosis of opium's ultimately stupefying effects.[9]

Keats cannot decide if he is in a "state" of "languor" or "Laziness", both of which had medical import: John Jones, in *Mysteries of Opium Reveal'd* (1700), cautioned that "a long and lavish Use of Opium"

[8] Fellow of the Royal College of Physicians; George I's physician-in-ordinary.

[9] Darwin's later medical text, *Zoonomia* (1974–96), liberally prescribed opium for hundreds of ailments.

means "the *sensitive soul* becomes lazy, listless, and averse to all *Actions*" (245); while Wilson described "the effects of opium [...] when immediately applied to the brain" as "impaired sensibility, languor, sleep" (1795, 101). The same terms had seen earlier use too. As early as 1640, Parkinson warned that "Opium [...] is such a medicine, [...] the continuall use whereof, bringeth [...] an insensiblenesse or stupefaction" (368). Burton's *Anatomy of Melancholy*, a favourite of Keats, advised readers that "Opium [...] stupefies" (2: 214). Crumpe describes initial "cheerfulness [...] subsiding into a state of pleasing languor, [...] stupor, and disinclination to motion" (1793, 45–46). Literary works straightforwardly propagated this medical diction: Aphra Behn, in the poem "Of Plants" (1689), for instance, could write casually of "Opium that binds the Nerves with Laziness" (1: 342); and in Mary Wollstonecraft's novel *Mary: A Fiction* (1788), "the opiate produced a kind of stupor" (179).

Keats' indolence was partially inspired literarily in "long[ing] after a stanza or two" of James Thomson's *Castle of Indolence* (1748), which describes "lulling" and "drowsy" "Indolence" as having "that slumbrous influence kest, | From poppies breathed" (Canto I, l.27, l.34, ll.21–22). At the period when opium was De Quincey's "elixir of pleasure", he too looked to Thomson's *Castle of Indolence* to express "happiness" (58–59). The term "indolence" was frequently used in medical texts to describe a person's state after consuming opium. Mead, for instance, averred cheerily: "They who take a moderate Dose of *Opium* [...] are so Transported with the pleasing Sense It induces" that they "injoy so perfect an Indolence and Quiet, that no Happiness in the World can surpass the Charms of this agreeable Extasie" (1702, 140–41); while military surgeon Alexander Grant, in *Observations on the Use of Opium* (1785), observed of his patients, "with some indolency took place" (15). Jones had explained: "It causes Indolence, or exemption from Pain" and "a Relaxation of all the sensible Parts of the Body" (1700, 23). These accounts accord with Keats' post-cricket-black-eye "relaxed" "indolence".

In his post-opium state of "indolence", Keats wrote in the letter of the dulling of both "pain" and "pleasure": "Pleasure has no show of enticement and pain no unbearable frown" (2: 78–79). Keats' medically inflected thinking in the letter directly works its way into his poetry (Keats 1958, 2: 79): "Ripe was the drowsy hour; / The blissful cloud of summer-indolence / Benumb'd my eyes; my pulse grew less and less; Pain had no sting, and pleasure's wreath no flower" ("Ode on Indolence" ll.15–18). The anatomical description of "Benumb'd" "eyes" and a "pulse" becoming "less and less" is redolent of medical texts. John Hill, in *Materia Medica* (1751), warned that opium brought "Dimness of the Eyes" and "a Slowness of the Pulse" (783). In November 1803, F. Weber noted during his own self-experimentation that he suffered from "heavy uneasy eyes; contracted pupils and dimness of sight" (436).

Pain-Relief as a Balm

In its status primarily as a medicine, opium was described as a remedy and relief from pain of almost all varieties, both physical and mental. Sydenham, in the mid-seventeenth century, lauded opium as a heaven-sent analgesic. Translating Sydenham's words, from Latin into English in 1696, John Pechey, himself a physician, rhapsodized: "[T]ruly I cannot here forbear mentioning, with Gratitude, that Omnipotent GOD, the Giver of all good Things, has not provided any other Remedy for the Relief of wretched Man, which is so able either to quell more Diseases, or more effectually to extirpate them, than Opiate Medicines" (140). Later, in the eighteenth century, Wilson assumes straightforwardly that readers will know from their own opium use that it brings "relief from pain" (1795, 3).

Johnson defined "balm" as "Any thing that sooth[e]s or mitigates pain", and many eighteenth-century poems praised opium's medicinal pain-quelling in the guise of being a "balm".[10] The anonymous "Laudanum", from the *London Magazine* (1735), refers to opium's "downy balm" which "quells the body's agonising pains" (l.22, l.6). "Orestes", in "A Sonnet to Opium", from the *European Magazine* (1796), apostrophised: "You can a balm infuse [...][.] You can the shaft of anguish quick remove" (ll.8–11). Maria Logan, who suffered prolonged illness, praised opium's balmy pain relief extensively in "To Opium" (1793): "Be mine the balm" to "still the throb of Pain", exhorting the "friendly balm" to "blunt each thorn" (ll.9–10, ll.23–27). In this light, Keats' "embalmed darkness", "as though" he had imbibed "some dull opiate", takes on enriched meaning in terms of being protected from experiencing worldly "sorrow" and "despairs", into a benumbed reverence "[t]o cease upon

[10] By 1742, it was known, through Charles Alston's experiments, that opium applied externally—i.e. in the manner we would today usually associate with the word *balm*—was ineffective (159).

the midnight with no pain" ("Ode to a Nightingale" l.43, l.2, l.3, l.27, l.28, l.56).

Sleep and Drowsiness

Following in Virgil's footsteps, Burton (1621) advised taking "opiats" to "procure" "sleepe" (2: 255). Many eighteenth-century physicians followed suit, prescribing opium as a sedative. Hill declared that physicians should "prescribe Opium principally as an Anodyne and Soporific", arguing that "The great Intent with which we give Opium is to procure Sleep" (1751, 782–83). Wilson observed that "Opium [...] produces [...] a tendency to sleep" (1795, 35–36). That it was a culturally widespread notion to prescribe opium as a sedative to the ill is evident from two consumptive characters in Wollstonecraft's *Mary*: "Ann soon fell asleep, as she had taken an opiate"; "Henry was certain he could not live long; all the rest he could obtain, was procured by opiates" (1788, 70; 176). Indeed, opium's tranquilising effects were well-known enough that Samuel Richardson, in *Clarissa* (1748–49), at the novel's crucial moment, could have his anti-hero Lovelace drug the heroine with "an opiate", inducing a state of "*somnivolence*", so he could rape her (897).

Unsurprisingly given his profession, Garth's poetical imagery of poppy-crowned people aligns with his contemporaries' accounts of opium's medical effects, principally slumber: "When the still Night, with peaceful Poppies crown'd", "Sleep shakes its downy Wings o're mortal Eyes" (1699, 53). Identical diction was used in private poetical writing too: Mary Allanson, a member of a Yorkshire gentry family near Ripon, never published any poetry, but she did write in her private verse notebook an "Address to Sleep March 1775" when tending to her gouty father: "Gentle Sleep thy pinions spread" "shake thy poppies thro the air" (l.1, l.24). That it was not just public figures who used imagery associating sleep and opium poppies suggests how widely this belief was embedded in cultural norms.

Keats also elides imagery of opium and sleep. In "Sleep and Poetry", for instance, he figures "[s]leep, quiet with his poppy crown" (l.348), and in "Eve of St Agnes" Madeline is with "poppied warmth of sleep oppress'd" (l.237). As in "To Autumn", Keats had associated sleepiness and drowsiness with opium, so "drowsy numbness" indicates an "opiate" acting on the body in "Ode to a Nightingale" (l.1, l.3). Many eighteenth-century poets had done likewise. William Harrison's "In Praise of Laudanum" (1714) opens with: "I Feel, O *Laudanum*, thy Power Divine, | And fall with Pleasure at thy slumb'ring Shrine" (ll.1–2); while Phanuel Bacon, in *Kite* (1722), states that a place of "sweet Repose" is where "the Drowsy Poppy grows" (4). The anonymous "Laudanum" begins: "The poppy [...] nodding waves its drowsy head" at "Somnus", the Roman God of sleep, and the author worries he/she can "feel thy drowsy influence in my verse" (1735, ll.1–4, l.52). George Crabbe, another physician-poet, in *Inebriety* (1775), described "drowsy peace" as "[s]leep [...] sheds her poppies" (8). By the end of the eighteenth century, the term "drowsiness" was an established medical term for opium's effects (Leigh 1786, 114; Crumpe 1793, 46). Indeed, William Buchan's standard household guide *Domestic Medicine* (1769), which went through twenty editions by 1800, asserted: "Too great a quality of opium generally occasions great drowsiness, with stupor" (527).

Dulling and Lulling

Medical texts explained opium's dulling psychopharmacological effects. Overuse of opium leads to a "dull, moapish, and heavy Disposition" (Jones 1700, 31), and "the mind becomes gradually dull and languid" (Crumpe 1793, 44). As well as dulling, opium also "lulls" its user: "It lulls, sooths, and (as it were) charms the Mind with Satisfaction, Acquiescence, Contention, Equanimity" (Jones 1700, 23).

Some poets of the period elaborate upon opium's "lulling" effects in more personal terms. Harrison wrote blithely: "Lull'd by thy Charms I' scape each anxious Thought, | And every thing but *Mira* is forgot" (1714, ll.3–4). "*Mira*" derives from the Latin for wonder—the same derivation as "miracle". In "Laudanum," opium "Lulls all the boiling passions to a calm", "And gently lulls into a soothing rest | The swelling sorrows of a troubled breast" (1735, l.23, ll.7–8). Anna Seward's "To the Poppy" (1810) describes opium as "lulling grief and pain" (l.12). Lady Caroline Lamb closes "Invocation to Sleep" wishing to "Lull, with thy poppy wreath, my soul to rest" (l.10).[11]

Like his predecessors, Keats used both "dull" and "lull" in association with opium. The "dull opiate" is recalled in the subsequent description of

[11] Lamb described "Invocation to Sleep" as "the first poem I ever wrote" (quoted in Douglass 2004, 27), and although it was published posthumously, in the *Keepsake* (1830), it is in keeping with pre-De Quinceyan conceptions of opium imagery.

"the dull brain [that] perplexes and retards" ("Ode to a Nightingale" l.3, l.34). Elsewhere, Keats describes "A breeze" which, having "through the dancing poppies stole", is "most softly lulling to my soul" ("Endymion" I:566–67). Written in mid-1819, at the same time as Keats' Great Odes, "Sonnet to Sleep" looks forward to when "thy poppy throws | Around my bed its lulling charities" (ll.7–8); the description of the "soft embalmer of the still midnight" (l.1) echoes the "embalmed darkness" of "Ode to a Nightingale" (l.43).

Oblivion and Forgetfulness

Garth describes opium as inducing "dull oblivion":

> No Passions interrupt his easie reign,
> No Problems puzzle his lethargick Brain.
> But dull oblivion guards his peaceful Bed,
> And lazy Fogs bedew his thoughtless Head. (1699, 5)

Likewise, Logan looks to opium to find an "oblivion" that "Relieves the weary'd mind" (ll.37–38). Seward notes "Kind dreams oblivious from thy juice proceed" (l.13). With the help of Sleep's "poppy wreath", Lamb sought to "Bind with Oblivion's veil these wakeful eyes" (l.10, l.3). Drawing explicitly on the classical allusion of Homer's Helen's opium drug, Thomson's "Indolence" drinks from "a fountain of Nepenthe rare: / Whence, as Dan Homer sings, huge pleasaunce grew, / And sweet oblivion of vile earthly care" (Canto I, ll.240–42). As Keats later would in "Ode to a Nightingale", Henrietta O'Neill, in "Ode to the Poppy" (1792), described opium's "soft Lethean power", a "balsam for a broken heart", in enabling one to forget (ll.45–46). Keats' "dull opiate" renders the narrative voice "as though" it "Lethe-wards had sunk" (ll.2–4); a voice that wants to "[f]ade far away, dissolve, and quite forget" earthly "weariness", "fever" and "fret" (ll.21–23). So accepted was opium's ability to lead to mental "oblivion" that William Hazlitt, writing in *The Spirit of the Age* (1825), could elide oblivion and opium in describing Coleridge as "swallowing doses of oblivion" (11: 34).

Metaphorical and Figurative Uses of Opium Imagery

The word opium thus became synonymous, symbolically, with sense-dulling. From the early seventeenth century, opium imagery was used figuratively as a literary metaphor divested of medical associations. In 1608, Thomas Morton could describe religious opponents as "Stupified with that *Opium* of implicit faith and blinde devotion" (33). An 1680s broadside poem, *Satyr Against Marriage*, attributed to the Earl of Rochester,[12] acerbically described marriage as "Opium to our Lustful Rage" (l.35). Pope considered "Dulness, the safe Opiate of the Mind, | The last kind Refuge weary Wit can find" ("Lines on Dulness" ll.1–2). Keats similarly uses opium imagery metaphorically to suggest mental dulling in having Albert wish to "smother" his "mind" from "the wrong I've done" with an "opiate for the conscience" ("Otho the Great" I.ii.155–58). In the same verse-letter relating the cricket incident, Keats employs opium imagery figuratively to describe how he will seek "revenge" on an advisory: "I will be opium to his vanity" (2: 90–91).

Writers who accused others of dullness often did so through opium imagery. John Dryden, in "MacFlecknoe" (1682), accused Thomas Shadwell, an opium-eater who subsequently died from overdosing, of being the Prince of Dullness, whose "Temples last with Poppies were o'er-spread" (l.126). Pope, in the *Dunciad* (1743), adopts the same imagery as Dryden: "And Shadwell nods the poppy on his brows" (III: 22). Pope used opium imagery to describe bad poetry in *Peri Bathos* (1728): metaphorically, "*opiates* of *poesy*" produce a "*tranquillity of mind*" which is a "*method* [...] used by our authors in *managing* the *passions*" because it "is a better design to promote sleep than madness" (213). Pope's *Dunciad*'s Goddess Dulness damningly has opium in her armoury: In "her own Guild-hall: | Here stood her Opium [...] The Goddess then o'er his anointed head, | With mystic words, the sacred Opium shed" so that "He sleeps among the dull" (I: 270–94). William Dodd took up Dryden's image, in the *Gentleman's Magazine* (1761), to criticise another poet, whose "insipidity of song" should be rewarded with "A wreath of *poppies* round thy temples" (ll.12–14). When Sir Robert Godschall petitioned the House of Commons in January 1742, Horace Walpole joked

[12] Though it may have been written by John Oldham or William Wycherley (Wilmot 1999, 490).

that his speech "proves so dull, one would think he chewed opium" (1: 163). By the eighteenth century, the image of opium as dulling was rooted in common parlance and image-making, beyond but coinciding with medical discourse.

So established in cultural convention was the imagery outlined above, within medical and literary genres alike—even despite the influence of prevalent alternative theories such as Brunonianism—that neoclassical conceptions of opium's effects persisted into Victorian accounts. George Eliot, for instance, writing well into the nineteenth century in *Silas Marner* (1861), used precisely the same diction. Eliot describes the "wretched unbenumbed consciousness" of Molly Farren without her "one comforter" "Opium": her "indolence" is assumed; she seeks "oblivion" in the drug, and once taken "she walked always more and more drowsily", is overcome by "weariness" and has "the longing to lie down and sleep", so much so that she falls into a "complete torpor" in a snowstorm and freezes to death (215–16).

Keats' poetry, like Eliot's conception in this episode, looks back to the neoclassical period's collective acceptance of opium's sedating and anaesthetising qualities, of both mind and body.[13] Following on from numerous neoclassical texts, Keats straightforwardly, at times, used opium imagery in conjunction with sleep: "Sleep", the "Low murmurer of tender lullabies", is "Wreather of poppy buds" ("Sleep and Poetry" ll.11–14). Elsewhere, Keats alludes to opium's dulling properties, and to the "drowsy numbness" one might experience having "emptied some dull opiate" into one's veins ("Ode to a Nightingale" ll.1–3). Keats had, in fact, amended this phrase by adding "drowsy", making it more akin to medical descriptions; he had initially written "painful numbness" (Hebron 2009, 137).[14] Opium can also lead "Lethe-wards", to forgetfulness ("Ode to a Nightingale" l.4). As indicated above, medical terms and imagery, common to the era, are manifested in Keats' poetry: with a "Dose" of "poppies" in early drafts of "To Autumn" (Hebron 2009, 143), and the description that his "pulse grew less and less" in "The blissful cloud" of post-opium "indolence" ("Ode on Indolence" l.15–17).

[13] N.B. the term *anaesthetic* itself did not materialise until the mid-nineteenth century.

[14] The poem's opening line, "My heart aches, and a drowsy numbness pains", was originally "My Heart aches and a painful numbness falls".

Scholars have seen De Quincey as exemplifying a Zeitgeist, according to which opium was an exotic means through which to attain mental liberation and concomitant imaginative spirit (Lefebure 1986, 51). The texts examined here, however, show that De Quincey's perception was far from medical and/or literary hegemony. Most long-eighteenth-century writing reflected the dulling, rather than transcendental or stimulating, effects of opium-taking as rooted in quotidian experience.[15] Standard imagery within long-eighteenth-century literature represented opium in line with medical neoclassical consensus: as inhibiting perception, rather than unshackling the mind. It was literature in which, as medical texts agreed, a dose of poppies left one drowsed.

References

Ackerknecht, Erwin. 2016. Medicine in the Seventeenth Century. In *Short History of Medicine*. Baltimore: Johns Hopkins University Press.

Alford, Henry, (trans.) 1861. *Odyssey*. London.

Allanson, Mary. Address to Sleep March 1775. Bodleian Library, Ms Eng Poet f.28, 17.

Alston, Charles. 1742. A Dissertation on Opium. *Medical Essays and Observations* 5 (1): 110–76.

Anonymous. 1735. Laudanum. *London Magazine* 4: 328–29.

Austen, Jane. 1995. *Letters*, ed. Deirdre Le Faye. Oxford: Oxford University Press.

Babington, William, and James Curry. 1811. *Outlines of a Course of Lectures on the Practice of Medicine, as Delivered in the Medical School of Guy's Hospital*. London: T. Bensley.

Bacon, Phanuel. 1722. *Kite*. Oxford: L. Lichfield.

Bate, Walter Jackson. 2014. *The Stylistic Development of Keats*. London: Routledge.

[15] Some eighteenth-century poets had, in fact, bucked the trend by treating opium as a creative agent. Thomas Warton, for instance, "stealing sleep", in *Pleasures of Melancholy* (1747), desired soaring dream-visions: "opiate dews" will "mystic visions send" "as Spenser saw" and "Milton knew" (7). Warton, ironically, differs from Spenser's and Milton's own opium-as-stupefying imagery mentioned above, but foreshadows De Quincey in consuming opium in order to unleash "his creative mind" and to be taken "thro' bewild'ring Fancy's magic maze" (7). But this was unusual. Indeed, in *Confessions*, De Quincey cites Shadwell (butt of Dryden's mockery) as the only poet he can "remember" who had "eaten opium" in order to procure "splendid dreams" (1998, 72).

Behn, Aphra. 1992–1996. *Works*, 7 vols., ed. Janet Todd. London: William Pickering.

Berridge, Virginia, and Griffith Edwards. 1987. *Opium and the People: Opiate Use in Nineteenth-Century England*. New Haven: Yale University Press.

Boon, Marcus. 2002. *Road of Excess: A History of Writers on Drugs*. Cambridge: Harvard University Press.

Beresford, James, (trans.) 1794. *Æneid of Virgil*. London.

Brown, Charles Armitage. 1937. *Life of John Keats*, ed. Dorothy Hyde Bodurtha and Willard Bissell Pope. London: Oxford University Press.

Brown, John. 1795. *Elements of Medicine*, 2 vols., trans. Thomas Beddoes. London.

Brown, Thomas. 1715. Translation from Horace of Mollis Inertia, February 1685. In *Fourth and Last of the Works*. London: Sam Briscoe.

Buchan, William. 1769. *Domestic Medicine; Or, the Family Physician*. Edinburgh: Balfour, Auld, and Smellie.

Buckley, Theodore Alois, (trans.) 1851. *Odyssey*. London.

Burton, Robert. 1989–2000. *Anatomy of Melancholy*, 6 vols., ed. Thomas Faulkner, Nicolas Kiessling, Rhonda Blair, J.B. Bamborough, and Martin Dodsworth. Oxford: Clarendon Press.

Byron, Lord George Gordon. 1973–1994. *Byron's Letters and Journals*, 13 vols., ed. Leslie Marchand. London: John Murray.

Chapman, George, (trans.) 1616. *Whole Works of Homer*. London: Richard Field.

Charvet, Pierre-Alexandre. 1826. *De l'action comparée de l'opium, et de ses principes constituans sur l'économie animale*. Paris: F. G. Lerrault.

Chaucer, Geoffrey. 2008. *Riverside Chaucer*, ed. Larry Benson. Oxford: Oxford University Press.

Cooper, Thomas. 1766. *Compendium of Midwifery; Or Pharmacopœia Obstetricaria*. London.

Cowper, William, (trans.) 1791. *Iliad and Odyssey of Homer*, 2 vols. London.

Crabbe, George. 1775. *Inebriety*. Ipswich.

Crumpe, Samuel. 1793. *Inquiry into the Nature and Properties of Opium*. London.

Culpeper, Nicholas. 1649. *Physicall Directory; Or, a Translation of the London Dispensatory Made by the Colledge of Physitians*. London.

Darwin, Erasmus. 1789. *Botanic Garden, Part II. Containing the Loves of the Plants*. Lichfield.

Darwin, Erasmus. 1794–96. *Zoonomia*. London.

Davenport-Hines, Richard. 2004. *Pursuit of Oblivion: A Global History of Narcotics*. New York: W. W. Norton.

De Quincey, Thomas. 1998. *Confessions of an English Opium-Eater*, ed. Grevel Lindop. Oxford: Oxford University Press.

Dodd, William. 1761. Verses Occasioned by Reading a Merely Descriptive Poem, Called Barham-Down. *Gentleman's Magazine* 31: 595.

Douglass, Paul. 2004. *Lady Caroline Lamb: A Biography*. Basingstoke: Palgrave Macmillan.

Dryden, John. 1956–1989. *Works*, 20 vols. Berkeley, Los Angeles, and London: University of California Press.

Edgeworth, Maria. 1999. *Belinda*, ed. Kathryn Kirkpatrick. Oxford: Oxford University Press.

Eliot, George. 1861. *Silas Marner: The Weaver of Raveloe*. Edinburgh.

Emerson, Oliver Farrar. 1919. Chaucer's 'Opie of Thebes Fyn'. *Modern Philology* 17 (5): 287–91.

Fairclough, H. Rushton, and G. P. Goold, (trans.) 2014. *Virgil*, 2 vols. Cambridge, MA: Harvard University Press.

Freeman, Hannah Cowles. 2012. Opium Use and Romantic Women's Poetry. *South Central Review* 29 (1-2): 1–20.

Garth, Samuel. 1699. *Dispensary*. London.

Ghosh, Hrileena, and Nicholas Roe. 2017. Formative Years and Medical Training. In *John Keats in Context*, ed. Michael O'Neill, 19–27. Cambridge: Cambridge University Press.

Goellnicht, Donald. 1984. *Poet-Physician: Keats and Medical Science*. Pittsburgh: University of Pittsburgh Press.

Grant, Alexander. 1785. *Observations on the Use of Opium*. London.

Hamilton, Robert. 1790. *Practical Hints on Opium*. Ipswich.

Harrison, William. 1714. In Praise of Laudanum. In *Poetical Miscellanies*, ed. Richard Steele, 245. London.

Hayter, Alethea. 1968. *Opium and the Romantic Imagination*. London: Faber.

Hazlitt, William. 1930–1934. *Complete Works*, 21 vols., ed. P.P. Howe. London: J. M. Dent.

Hebron, Stephen. 2009. *John Keats: A Poet and His Manuscripts*. London: British Library.

Hill, John. 1751. Opium. In *History of the Materia Medica*, 779–84. London.

Hobbes, Thomas, (trans.) 1675. *Homer's Odysses*. London.

Holloway, S.W.F. 1995. Regulation of the Supply of Drugs in Britain Before 1868. In *Drugs and Narcotics in History*, ed. Roy Porter and Mikuláš Teich, 77–96. Cambridge: Cambridge University Press.

Hutchinson, Sara. 1954. *Letters*, ed. Kathleen Coburn. London: Routledge and Kegan Paul.

Johnson, Samuel. 1755. *Dictionary of the English Language*, 2 vols. London.

Jones, John. 1700. *Mysteries of Opium Reveal'd*. London.

Keats, John. 1958. *Letters*, 2 vols., ed. Hyder Edward Rollins. Cambridge, MA: Harvard University Press.

———. 1978. *Poems*, ed. Jack Stillinger. London: Heinemann.

Kirkland, Thomas. 1780. Short Essay on the Use of Opium. In *Thoughts on Amputation*, 58–70. London.

Lamb, Caroline. 1830. Invocation to Sleep. In *Keepsake*, ed. Frederic Mansel Reynolds, 202. London.

Leask, Nigel. 1992. 'Murdering One's Double': Thomas De Quincey and S. T. Coleridge. Autobiography, Opium and Empire in *Confessions of an English Opium Eater* and *Biographia Literaria*. In *British Romanticism Writers and the East: Anxieties of Empire,* 170–228. Cambridge: Cambridge University Press.

Lefebure, Molly. 1986. Consolations in Opium: The Expanding Universe of Coleridge, Humphrey [*sic*] Davy and *The Recluse*. *Wordsworth Circle* 17 (2): 51–60.

Leigh, John. 1786. *Experimental Inquiry into the Properties of Opium*. Edinburgh.

Logan, Maria. 1793. To Opium. In *Poems on Several Occasions,* 17–21. York.

Maehle, Andreas-Holger. 1995. Pharmacological Experimentation with Opium in the Eighteenth Century. In *Drugs and Narcotics in History*, ed. Roy Porter and Mikuláš Teich, 52–76. Cambridge: Cambridge University Press.

———. 1999. *Drugs on Trial: Experimental Pharmacology and Therapeutic Innovation in the Eighteenth Century*. Amsterdam: Rodopi.

Mead, Richard. 1702. Essay IV 'Of Opium'. In *Mechanical Account of Poisons,* 131–48. London.

Milligan, Barry. 1995. *Pleasures and Pains: Opium and the Orient in Nineteenth-Century British Culture*. Charlottesville: University Press of Virginia.

———. 2007. Brunonianism, Radicalism, and 'The Pleasures of Opium'. In *Thomas De Quincey: New Theoretical and Critical Directions*, ed. Robert Morrison and Daniel Sanjiv Roberts, 45–61. London: Routledge.

Milton, John. 1997. *Complete Shorter Poems*, ed. John Carey. Harlow: Pearson Longman.

Monro, Alexander. 1771. Attempt to Determine by Experiments, How Far Some of the Most Powerful Medicines, Viz. Opium, Ardent Spirits, and Essential Oils, Affect Animals. In *Essays and Observations, Physical and Literary*, vol. 3, 292–365. Edinburgh.

Morton, Thomas. 1608. *Preamble Unto an Incounter with P.R. the Author of the Deceitful Treatise of Mitigation: Concerning the Romish Doctrine*. London.

Murray, Augustus Taber, and George Dimock, (trans.) 2014. *Odyssey*, 2 vols. Cambridge, MA: Harvard University Press.

'O'. 1783–1789. Review of *Observations on the Use of Opium* by Alexander Grant. *English Review*, 14 vols., 5:451–54. London.

Ober, William. 1968. Drowsed with the Fume of Poppies: Opium and John Keats. *Bulletin of the New York Academy of Medicine* 44 (7): 862–81.

Ogilby, John, (trans.) 1665. *Homer, His Odysses Translated*. London.

O'Neill, Henrietta. 1792. Ode to the Poppy. *Desmond*, 3 vols., ed. Charlotte Smith, 3:165–66. London.
'Orestes'. 1796. Sonnet to Opium; Celebrating its Virtues. *European Magazine* 30: 54.
Ovid. 2014. *Heroides; Amores*. (trans.) G. P. Goold and Grant Showerman. Cambridge, MA: Harvard University Press.
Parkinson, John. 1640. *Theatrum Botanicum*. London.
Pepys, Samuel. 2016. *Diary*, 11 vols., ed. Robert Latham and William Matthews. London: HarperCollins.
Pope, Alexander. 1961–1969. *Twickenham Edition of the Poems*, 11 vols., ed. John Butt. London: Methuen.
———. 2008. Peri Bathos; or, The Art of Sinking in Poetry. In *Major Works*, ed. Pat Rogers, 195–239. Oxford: Oxford University Press.
Richardson, Samuel. 1985. *Clarissa*, ed. Angus Ross. Harmondsworth: Penguin.
Roe, Nicholas. 2012. *Keats: A New Life*. New Haven: Yale University Press.
Scarborough, John. 1995. The Opium Poppy in Hellenistic and Roman Medicine. In *Drugs and Narcotics in History*, ed. Roy Porter and Mikuláš Teich, 4–23. Cambridge: Cambridge University Press.
Schmitt, Cannon. 2002. Narrating National Addictions: De Quincey, Opium, and Tea. In *High Anxieties: Cultural Studies in Addiction*, ed. Janet Farrell Brodie and Marc Redfield, 63–84. Berkeley, Los Angeles, and London: University of California Press.
Seward, Anna. 1810. To the Poppy. *Poetical Works*, 3 vols., ed. Walter Scott, 3:192. Edinburgh.
Shakespeare, William. 2008. *Royal Shakespeare Company Complete Works*, ed. Jonathan Bate and Eric Rasmussen. Basingstoke: Macmillan.
Shewring, Walter, (trans.) 1998. *Odyssey*. Oxford: Oxford University Press.
Siegel, Ronald, and Ada Hirschman. 1983. Charvet and the First Psychopharmacological Studies on Opium. *Journal of Psychoactive Drugs* 15 (4): 323–29.
Sotheby, William, (trans.) 1834. *Iliad and Odyssey of Homer*, 4 vols. London.
Spenser, Edmund. 2013. *Faerie Queene*, 2 vols., ed. J.C. Smith. Oxford: Oxford University Press.
Sydenham, Thomas. 1696. *Whole Works*, trans. John Pechey. London.
Thomson, James. 2014. *Liberty, Castle of Indolence, and Other Poems*, ed. James Sambrook. Oxford: Oxford University Press.
Trapp, Joseph, (trans.) 1731. *Works of Virgil*, 3 vols. London.
Walpole, Horace. 1903–1918. *Letters*, 16 vols., ed. Helen Toynbee. Oxford: Clarendon Press.
Warton, Thomas. 1747. *Pleasures of Melancholy*. London.
Weber, F. 1803. Observations on the Effects of Opium on the Human Body. *Medical and Physical Journal* 10 (57): 435–43.
Whytt, Robert. 1768. *Works*. Edinburgh.

Wilmot, John [Earl of Rochester]. 1999. *Works*. Ed. Harold Love. Oxford: Oxford University Press.

Wilson, Alexander Philip. 1795. *Experimental Essay, on the Manner in which Opium Acts on the Living Animal Body*. Edinburgh.

Wollstonecraft, Mary. 1788. *Mary: A Fiction*. London.

Wortley Montagu, Lady Mary. 1977. *Essays and Poems*, ed. Robert Halsband and Isobel Grundy. Oxford: Clarendon Press.

Wu, Duncan. 2015. *30 Great Myths about the Romantics*. Chichester: Wiley.

Young, George. 1753. *Treatise on Opium, Founded Upon Practical Observations*. London.

CHAPTER 3

"Grief's Comforter, Joy's Guardian, Good King Poppy!": Opium and Victorian Poetry

Irmtraud Huber

> I duly visited the tomb of the poet Coleridge, which lies in the cloisters under the chapter, the lodge-keeper [...] appeared both willing and anxious to let me know that Muster [*sic*] Coleridge were a poet and took opium. From his tone, I judged that as far as his acquaintance with poets went, he considered opium-eating a natural consequence of the poetic temperament, and that the poet who couldn't take his opium like a man was rather a poor sort of creature. (Anonymous 1895, 290)

This humorous circumstantial conclusion to an article on the state of school music at Highgate, published in 1895, is suggestive of the extent to which poetry and opium-eating could be associated in the popular mind during Victoria's reign. The notoriety of Samuel Coleridge's opium habit, which became a matter of public debate shortly after his death in 1834, had a significant role to play in this, as had the immense influence of Thomas De Quincey's *Confessions of an English Opium-Eater*

I. Huber (✉)
Ludwig-Maximilians-Universität München, Munich, Germany
e-mail: Irmtraud.Huber@anglistik.uni-muenchen.de

N. Roxburgh and J. S. Henke (eds.), *Psychopharmacology in British Literature and Culture, 1780–1900*, Palgrave Studies in Literature, Science and Medicine,
https://doi.org/10.1007/978-3-030-53598-8_3

(1821), in which the author likens his opium-induced dreams to poetic visions and famously suggested: "We hear it reported of Dryden and of Fuseli, in modern times, that they thought proper to eat raw meat for the sake of obtaining splendid dreams: how much better for such a purpose to have eaten opium" (2013 [1821], 71). Nor did the Victorians lack further examples of poets with an actual or popularly alleged opium habit, like Elizabeth Barrett Browning, Francis Thompson or James Clarence Mangan. Meanwhile, the persistent influence of the commonplace is perhaps best suggested by the figure of Augustus Carmichael, the languid, opium-eating poet lounging in the sun in Virginia Woolf's *To the Lighthouse* (1927).

The question of to what extent opium intake might have had an influence on poetic expression has since been the matter of some critical debate, focusing mainly on the work of Romantic and early Victorian poets with a known opium habit (Abrams 1971 [1934]; Hayter 1968). More recently, however, historians have suggested that "[e]xcessive concentration on [De Quincey's and Coleridge's] spectacular histories of addiction [...] has tended to disguise the overlapping of addiction, social and medical usage [of opium] throughout middle-class society at the time" (Berridge 1999, 56). Virginia Berridge emphasises the common usage and wide distribution of the drug at the time and suggests that "opium was a simple part of life" (1999, 61). As a popular medicine, opium was used across all strata of society, and the evidence Berridge collects in her study seems to suggest that most Victorians would have taken at least small dosages of some kind of opiate at some point in their lives.

It is therefore not surprising that opium features in a number of Victorian poems. Beyond noting this fact, in this paper I explore an interrelation between poetry and drug effects which entangles questions of form and content, reference to the drug and imitation of its effects. I also maintain that a focus on Victorian poetry adds to and complicates the picture that emerges from the perhaps more familiar literary representations of the drug to be found in Romantic poetry or Victorian fiction. Moreover, I suggest that there may be a link between the cultural shift which increasingly privileged narrative fiction over poetry over the course of the nineteenth century on the one hand and the rise of a discourse of addiction on the other. I locate this link in the way in which poetry and narrative fiction, and the nineteenth-century realist novel in particular, are invested to a different degree in the idea of the autonomous, rational individual, which is in control of his or her actions and desires. Thus, my investigation of the relation between opium and poetry serves more broadly to highlight underlying cultural and ideological commitments of particular literary forms.

Recent critical interest has turned in particular towards the significance of opium as a commodity deeply embedded in colonial power struggles and xenophobic anxieties, exploring the way in which it serves both "as a metaphor for the dynamics of Anglo-Oriental exchange and as an actual commodity caught up in that exchange" (Milligan 1995, 29). If the earlier critical interest in opium and literature tended to focus on a small group of (mainly) Romantic poets,[1] such postcolonial approaches find their material more often in prose and fiction: apart from the omnipresent De Quincey, such discussions frequently feature depictions of opium dens by Charles Dickens, Oscar Wilde and Arthur Conan Doyle, and the drug's role in sensation novels like Wilkie Collins' *The Moonstone* (1868) (Milligan 1995, 69–101; Polsky 2015; Roth 2002). Whereas the earlier work of M. H. Abrams (1971 [1934]) and Alethea Hayter (1968) discusses opium as an at least potentially creative and inspiring source for the work of the poets, albeit with highly pernicious consequences, a focus on its role in late nineteenth-century fiction generally paints a thoroughly sinister picture, revealing fears about "China invad[ing] England through the East End opium den and proceed[ing] to turn its citizens into addicts" (Roth 2002, 91).

This development seems to accord nicely with a considerable change in the general attitude towards opium following the 1848 Pharmacy Act and the endeavours of the anti-opium movement, combined with an increasing awareness of the drug's addictive properties (Berridge 1999, 173–93, 225–31). It might seem, at first glance, that literature merely reflects this turn from a largely positive popular attitude towards the drug in the early nineteenth century, which valued it for its medical uses and generally tolerated its recreational use as lamentable but harmless, to a growing awareness of the problem of drug abuse in a progressive society which had more knowledge about the adverse effects of the drug and

[1] In an attempt to broaden this focus, Hannah Cowles Freeman has suggested that the work of Romantic women writers like Anna Seward, Henrietta O'Neill, Sara Coleridge and Maria Logan "offer[s] a counter discourse to the traditional opium canon", as their poems "incorporate opium use into the domestic hearth, its medicinal effects aiding in their wifely and motherly duties" (2012, 1).

the dangers of drug habituation. However, as I will demonstrate, this apparently straightforward critical narrative of progress turns out to be rather too facile, and a focus on Victorian poetry is one way to complicate matters. My reading of Victorian poetry will press two points:

My first contention is that, as a Romantic heritage, the association between poetry and opium remains strong for the Victorians, but is also consistently troubled. Most significantly, I argue for a shift in the quality of the association. There was clearly a strong sense of the pernicious effects of opium and the dire consequences of the habit, both among the Romantics themselves and among the Victorian commentators on Coleridge's and De Quincey's life. Nonetheless, poems like Coleridge's "Kubla Khan" or Keats' "Ode to the Nightingale" as well as De Quincey's *Confessions* suggested influentially that opium intake could be conducive to poetic vision. The Victorians, in contrast, developed the analogy on the basis of similar effects. Rather than inspiring poetry, opium and poetry share common properties. As one anonymous Victorian reviewer of a now forgotten poet put it in 1854: "In this hard age, we take poetry, like opium, to deaden our reason, to blunt our sorrows, erase our cares, to steep us in forgetfulness, and to enchant us with delicious dreams" (Anonymous 1854, 275).

The full implications of these observations emerge when one calls to mind that, according to current knowledge, opiates have neither a directly hallucinogenic effect nor are they particularly soporific; that is, their intake generally leads neither to a heightening of the senses and illusionary visions, nor do they induce sleep.[2] According to Elizabeth Martin's entry in the *Oxford Concise Medical Dictionary*, opiates are analgesics, that is, painkillers. They "depress the central nervous system: they relieve pain, induce feelings of euphoria, suppress coughing, and stimulate vomiting" and their "side-effects include drowsiness, nausea and vomiting, constipation, and depression of breathing" (Martin 2015). We observe euphoria, but no imaginative flights; drowsiness, but not sleep. From this perspective, the inspirational opium dream seems to be almost as much a fiction as the sinister opium den.

[2] Eellan Sivanesan, Melvin C. Gitlin and Keith A. Candiotti note that "[o]pioid hallucination is an *uncommon* yet significant adverse effect of opioid treatment" (2016, 836 emphasis added). A recent study of adverse side effects of morphine usage in emergency settings does identify drowsiness as a possible side-effect (though by no means the main drug effect) and does not mention hallucinations at all (Bounes et al. 2017).

It is possible, of course, that historical differences in the kind of opium intake can account for different effects.[3] Contemporary medical texts, in any case, disagreed about whether opium should be considered what they called a stimulant or a narcotic. Gavin Budge notes among Victorian sources a common "characterization of alcohol and opium as 'stimulants', rather than the narcotics or depressants as which nowadays we more readily regard them" and suggest that this "reflect[s] the terminology of Brunonian medicine" (2013, 56). Brunonianism, based on doctrines developed by John Brown in the late eighteenth century, held that all disease was an effect of over- or under-stimulation, and opium was Brown's favoured stimulant. This terminology, however, would prove misleading. Writing in 1864, Francis Edmund Anstie laments the common "confusion and uncertainty in the application of [...] the words 'Narcotic' and 'Stimulant'" (Anstie 1864, 1), a confusion deriving partly from the fact that Brown did not make this distinction. "The author of Brunonism", Anstie claims, "had stated that there was, in truth, no such thing as direct *sedative action*; and that all the so-called sedative affections were produced by the use of a weaker stimulus than usual" (1864, 46). Anstie, for his part, considers opium to have both stimulating and narcotic properties, depending on the dose (1864, 117–18), but he also notes that the effects of the drug still remain uncertain, for lack of "precise data" (1864, 79). As a still unfamiliar "oriental" drug with uncertain effects, opium offered a foil for idealising as well as demonising projections. Poetic representations of opium, I argue, may therefore say at least as much about attitudes towards poetry as they say about the effects of drug action.

My second contention is that genre and form play an important role in literary attitudes towards opium. To some extent, I would suggest, the dual nature of opium, beneficial/medicinal on the one hand and addictive on the other, can be traced in the above-mentioned contrast in critical attention to poetry and prose fiction. Susan Zieger has suggested that "[t]he questions that addiction begs – 'How did it start?' 'When did it all begin?' and, bathetically, 'How did it come to this?' – are questions that generate narrative" (Zieger 2008, 3). But the converse logic also applies: narrative engagements with opium use, at least ones

[3] Hayter warns against historical anachronism in judging drug effects, reminding readers that "[w]hat the modern addict takes is different in itself, and differently administered; and he takes it in a different climate of opinion" (1968, 37).

indebted to the aesthetics of the novel, tend to generate stories of addiction. One reason for this, I suggest, may lie in the historical development of the novel as a distinct literary genre and the importance of individualism for this history. Ian Watt argued in his influential study *The Rise of the Novel* that what distinguishes the novel is that its "primary criterion was truth to individual experience – individual experience which is always unique" (2001 [1957], 13; see also 60–62) and Nancy Armstrong suggests that the essential ideological core of the genre is "the presupposition that novels think like individuals about the difficulties of fulfilling oneself as an individual under specific cultural historical conditions" (2005, 10). Unique individual experience and fulfilment as an individual are, however, threatened by addiction, in which "[t]he subject of desire has been produced by the product, even though the product is nothing more than a placeholder for desire" (Brodie and Redfield 2002, 8; see also Zieger 2008, 6–11). From this perspective, addiction threatens the basic premise of the aesthetic project of the novel, eroding it, as it were, from the inside. At the same time, Peter Brooks has influentially argued for narrative desire as the fundamental driving force of plot, with the heroes of the nineteenth-century novel as prime examples for what he calls "desiring machines" (1984, 39). There is a thin line between perpetuated desire and addiction, made particularly visible in the serial form in which much nineteenth-century fiction was published, which could itself potentially develop addictive properties.[4] Addiction anxieties as they are often expressed in fiction may thus also obliquely refer to anxieties about fiction's own potentially addictive form. From this perspective, it is not a coincidence that the concept of addiction gains traction in a cultural climate which has been nourished on notions of autonomy, individuality and selfhood promulgated by the novel. This is the climate in which addiction, in Eve Sedgwick's words, can be understood to "reside [...] only in the structure of will that is always somehow insufficiently free, a choice whose voluntarity is insufficiently pure" (1993, 132).

The association of poetry with opium, however, generally follows a different logic, rooted in a Romantic ideal of poetic vision best embodied in the intense and condensed expression of the lyric. In depicting opium, nineteenth-century fiction typically traces circumstances, causes

[4] Susan Zieger suggests that the contemporary hype around the Sherlock Holmes stories can be seen as an early form of media addiction (2014, 25–26). See also the essay in this volume by Sarah Frühwirth.

and effects of addiction, while poetry more often explores the altered states of mind the drug offers, sidelining its long-term addictive effects. If one follows the argument made about the novel's commitment to the autonomous individual above, one might here point to the universalising tendencies of lyric subjectivity which exceed the purely individual. Mutlu Konuk Blasing suggests that "there is no 'individual' in the lyric in any ordinary sense of the term" (2007, 4). Subjectivity is obviously of central importance to the lyric, but not the unique, autonomous individual which is at the core of the nineteenth-century novel. The "I" in lyric poetry is usually little individualised, typically allowing the reader, at least to some extent, "to utter [the poem] as one's own words, not as the words of another" (Vendler 2014, 133). Moreover, the lyric as a contemplative genre is rarely concerned with action. The lyric "I" is thus generally porous and adaptable and claims little agency. Poetry more generally was valued in the nineteenth century particularly for its imaginative vision, its vivid imagery and for a sense of heightened aesthetic (and synaesthetic) sensibilities. It was supposed to give pleasure by addressing the emotions, to stimulate the imagination and to offer an escape from reality. It typically explores momentary somatic perception, not sequences of cause and effect like narrative prose does. Purely in terms of form, the novel and poetry (lyric poetry in particular) thus show affinities towards different aspects of drug effects. Drastically simplified one could say that lyric poetry relates to the brief drug high, narrative prose to the never-ending story of addiction. Nonetheless, as we will see, in Victorian poetry the spectre of addiction frequently rears its head, as it were, at the limits of the poetic text.

Victorian Opiate Poetry: Poetic Drug Effects

The Victorian association between poetry and opium on the basis of known drug effects derives from a cultural attitude towards poetry which values its consolatory power in times of grief. Drawing attention to the prevalence of melancholy and mourning in Victorian poetry, Valentine Cunningham claims that "Victorian poems are multiplied scenes, extended cries, of desolation" (2011, 324) and suggests that the production of elegies is fuelled by what he calls a "prophylaxis of melancholy – melancholy as a medicinal condom", "a steady intake of melancholic stuff as the only prophylactic against being utterly overwhelmed by morbidity and mortality" (2011, 324, 343). Poetry is the narcotic which is administered to help bear the desolation of existence.

The popularity of such a narcotic model of poetry owes much to Alfred Tennyson's quintessential poem of consolation, *In Memoriam A.H.H.* (1850).[5] In section V of this poem, Tennyson draws a parallel between the effect of poetry, or, more specifically, the rhythmical language of rhyme and metre, and the narcotic effects of opiates:

> I sometimes hold it half a sin
> To put in words the grief I feel;
> For words, like Nature, half reveal
> And half conceal the Soul within.
>
> But, for the unquiet heart and brain,
> A use in measured language lies;
> The sad mechanic exercise,
> Like dull narcotics, numbing pain. (ll. 1–8)

This is no celebration of drug-induced poetic vision. Instead, the drug is likened to poetry on the basis of their common property of offering escape and solace, but the escape offered is not one of heightened senses, but the numbing of pain. There is a hint of the addictive properties of the drug, of the structure of compulsory repetition which it shares with the "measured language" of verse. As Roger Platizky puts it, "[l]ike 'the sad mechanic exercise' of writing stanza after stanza in a compulsive *abba* rhyme, the drugs must be repeated as the temporary effect of the numbness wears off" (2002, 211–12). However, instead of an addictive spiral of deterioration, the relentless repetitions of the poem's metrical structure provide order and support even in the darkest moments of doubt, grief and despair. Nonetheless, the "dull narcotics" of metrical language are guilt-ridden. It seems "half a sin" to seek escape or reprieve from the stark reality of the pain, to hide it in words. Poetic production, like opium intake, becomes a guilty habit, and the stanzas are ambivalent about whether this habit is excusable and necessary in face of the immensity of pain, or cowardly and treacherous because it conceals and shelters the soul from the world. Dream visions are associated with opiates in a

[5] Full-text versions of all the poems discussed in this paper can be found in the Chadwyck-Healey collection of English Poetry, managed by *ProQuest*.

later section of *In Memoriam*, but again the drug is called upon not to heighten the senses but to dull pain. Dreams bring "A night-long Present of the Past" (LXXI, l. 3) in which the speaker and his dead friend are reunited, but only opiates "trebly strong" could "Drug down the blindfold sense of wrong" (LXXI, l. 6–7), thus making the illusion whole. An analogy between poetry and opiates continues to haunt this section of the poem, in particular if it is read in the light of the resolution of *In Memoriam*, in which the poem itself has finally achieved a subdual of the sense of wrong in a joyful imaginary reunion with the dead friend: "Far off thou art, but ever nigh; / I have thee still, and I rejoice; / I prosper, circled with thy voice; / I shall not lose thee tho' I die" (CXXX, l. 13–16). Poetry finally succeeds in accomplishing what the trebly strong opiate was called on to do.

Strikingly, this analogy between poetry and a narcotic drug like opium is even borne out in *In Memoriam*'s reception. The poem was celebrated for its powers of consolation, with Queen Victoria as its most famous proselyte. As Kristie Blair comments, "[r]eading *In Memoriam* is discussed [by contemporaries] in terms that suggest it could help ease bodily suffering" (2001, 247). It thus serves as a prime example for a model of poetry that stresses its soothing effects: "poetry as a means to induce peace and tranquillity by calming body and mind and to provide 'healing relief' from dangerous emotion" (Blair 2001, 247). Poetry, in other words, serves as a beneficial analgesic. Such an attribution of physical drug effects to the reading of poetry is aligned with Brunonian thought. Budge draws attention to the Brunonian underpinnings of ideas developed in William Wordsworth's preface to the *Lyrical Ballads* (Budge 2013, 14). Wordsworth's aim to provide a poetic antidote to the overstimulation of modernity, a "kind of mild stimulant [...] that consequently [...] can have a healing effect on its readers" (Budge 2013, 62), is an important context for understanding the reactions to *In Memoriam* Kirstie Blair discusses. At the same time, there is a sense in which the seemingly endless repetitions of *In Memoriam* are excessive in their proclamation of grief. In the very escape and consolation the poem offers lies hidden a darker, perhaps even addictive effect, giving an "incentive to continue mourning well past the accepted bounds" (Blair 2001, 251). Blair notes that "[w]hile the form and message of Tennyson's poem were largely read as redemptive, this reading coexists with another interpretation of the poem as morbid, dwelling on grief and doubt" (2001, 251). Poetry is not only narcotic; it also perpetuates a grief which can itself come close to an addiction. As Cunningham suggests: "Like memories of the dead, writing in and about melancholy is very hard, if not impossible, to let go of" (2011, 343). Drawn out to the length of the 131 sections of *In Memoriam*, poetry's narcotic effect may apparently threaten to become addictive.

If opium's property to offer escape from pain and from a weary life is appreciated by the Victorians, the drug also becomes increasingly morally questionable in a cultural climate which values dedication to duty, useful productivity and steadfastness. These conflicts find iconic expression in another of Tennyson's major poems, the deeply ambiguous "The Lotos-Eaters" (1832), in which the mariners foreswear the ceaseless labour of "ever climbing up the climbing wave" (l. 95) and yearn for the "long rest or death, dark death, or dreamful ease" (l. 98) which the consummation of the lotus flower grants them, but do so at the price of relinquishing their responsibilities towards their families and even their very humanity. In such a context, the association between poetry and drugs (or the poet and the lotus eater) turns into a challenge to reconsider the poet's place in the world (O'Gorman 2004, 64). Tennyson's masterful engagement with the ambiguities involved in these issues has drawn critical attention (Armstrong 2013; Colman 2019; Platizky 2002; Stevenson 1982), and in what follows I prefer to explore less frequently discussed poems. It seems worth noting in passing, however, that "The Lotos-Eaters" is a poem about choice as much as about addiction. Though Adam Colman discusses the poem as an example of what he calls an "addiction aesthetic" (2019, 3), the rhetoric of the poem as well as the basis for the moral judgement on the mariners it invites depends on an assertion of will, rather than addictive dependence. "[W]e will no longer roam" (l. 45), the mariners sing, and "[w]e have had enough of action, and of motion we" (l. 150). The moral issue the poem raises concerns a wilful withdrawal from the world, not the kind of loss of control and agency implied by addiction.

With critical discussion of opium in Victorian poetry focusing primarily on "The Lotos-Eaters", the presence of similar issues in other poetry of the period has largely escaped attention. Francis Thompson's "The Poppy" (1891), for example, turns the flower into a complex cipher for the poet, for the content of his poetry (fickle love) and for this poetry itself, but in the poem's repeated semantic substitutions all of these associations become subtly ambivalent. Like the poppy among the corn, the poet is an idle dreamer among productive, "goodly men":

The sleep-flower sways in the wheat its head,
Heavy with dreams, as that with bread:
The goodly grain and the sun-flushed sleeper
The reaper reaps, and Time the reaper.

I hang 'mid men my needless head,
And my fruit is dreams, as theirs is bread:
The goodly men and the sun-hazed sleeper
Time shall reap, but after the reaper
The world shall glean of me, me the sleeper. (ll. 64–72)

The comparison between poet and poppy is in this case based on the power to provide dreams but also suggests a different similarity: their "needless" unproductivity. In analogy to the poppy, the poet becomes a dreamy weed, albeit a weed whose remains may turn out to outlive the corn. Both the weed's (and poet's) excessiveness and its potential to outlast the corn are further emphasised by the additional, excessive line and rhyme of this stanza. The comparison appears hardly complimentary in view of the morally dubious imagery with which the poppy flower is introduced at the beginning of the poem. Drunk, as it were, from the lushness of the poetic imagery developed in the first two stanzas, which pile imagery upon imagery, the poppy is likened to a drugged, lecherous gipsy:

Summer set lip to earth's bosom bare,
And left the flushed print in a poppy there:
Like a yawn of fire from the grass it came,
And the fanning wind puffed it to flapping flame.

With burnt mouth, red like a lion's, it drank
The blood of the sun as he slaughtered sank,
And dipped its cup in the purpurate shine
When the eastern conduits ran with wine.

Till it grew lethargied with fierce bliss,
And hot as a swinked gipsy is,
And drowsed in sleepy savageries,
With mouth wide a-pout for a sultry kiss. (ll. 1–12)

As the poet is likened to the poppy, poetry is analogous to a drug that brings dreamy lethargy but elicits moral condemnation. When the speaker is given a poppy flower as an impulsive gift from a beloved child, this gift of a "withering flower of dream" (l. 73) is bitterly appropriate, not only because the speaker deems the child's attachment evanescent and immature, but also because the contrast between the innocent child and the drugged, lecherous poppy-gipsy-poet suggests an impropriety in the attachment. Moreover, the poppy comes to stand for the frustrated erotic desire of the speaker, which is timelessly perpetuated in the repetitions of rhyme (and its ceaseless variation of repetitions is the most prominent and obvious rhetoric feature of the poem). The poppy, both flower and poem, is "Sheltered and shut in a nook of rhyme" (l. 75) by the poet, but while he evokes the topos of art outlasting life, the supposed purity of art is compromised by its association with the morally dubious poppy as well as by the similarity of its economy of desire to the ever-perpetuating "time-destroying repetitions" of addiction (Zieger 2008, 3). Moreover, the poetic visions the poppy grants are no more than "withered dreams": "Love! I fall into the claws of Time: / But lasts within a leavèd rhyme / All that the world of me esteems— / My withered dreams, my withered dreams" (ll. 60–80). While the dream-inspiring qualities of the poppy are central to this poem's imagery, this is a far cry from the opium-induced visionary inspirations of "Kubla Khan" and "Ode to a Nightingale". Instead of being a source of creativity, the creative power of the poet is questioned and undermined by his association with the poppy, whose opiate dreams are withered and withering at the same time.

In an even more sombre mood, the dreams James Thomson hopes to receive from opium intake in his celebration of suicide "Our Ladies of Death" (1860) are not visionary or inspirational, but evolve "Some ever-slumbrous, never-varying theme" (l. 181). In this poem, inspired by De Quincey's *Suspira de Profundis* (1845), Thomson envisages the escape offered by opium not as transcendent vision, but as the "restful rapture of the inviolate grave" (l. 175). In what reads like a morbid parody of a Wordsworthian union with nature, Thomson concludes this poem by eulogising death as an "abdication of my separate soul" (l. 193):

> One part of me shall feed a little worm,
> And it a bird on which a man may feed;
> One lime the mould, one nourish insect-sperm;
> One thrill sweet grass, one pulse in bitter weed;
> This swell a fruit, and that evolve in air;
> Another trickle to a springlet's lair,
> Another paint a daisy on the mead (ll. 204–10)

The drug, which offers a "divinely tranquil Death" (l. 223) to the poet who is "Weary of living isolated life / Weary of hoping hopes for ever vain, / Weary of struggling in all-sterile strife" (ll. 208–10), thus suggests a darkly literal realisation drug-induced poetic vision as the opium dream is replaced by an opium death.

The most elaborate and remarkable explicit poetic engagement with opium among the Victorians, however, is the genre-defying, book-length *King Poppy* by Robert Bulwer Lytton (also known under his pen-name Owen Meredith and not to be confused with his more famous father, the novelist Edward Bulwer Lytton). Published posthumously in 1892, *King Poppy* involves a curious reworking of the affinities that seemed to exist between poetry and opium for the Victorians. It is also a clear indication that even late-Victorian attitudes towards the drug and its effects were not unanimously negative or anxious. After all, the poppy is one of the poem's central heroes and is celebrated in the highest terms in the poem's conclusion:

> Reign for ever,
> Fate's master, yet man's ministering friend!
> Guide of stray'd Love's lone footsteps to the goal
> That, lost on earth, is in thy realm recover'd!
> Feaster of famisht hearts, rebuilder bright
> Of ruin'd fortunes, pain's victorious foe,
> Grief's comforter, Joy's guardian, good King Poppy! (ll. 5305–11)

None of the contemporary reviewers of the poem found such unreserved praise of opium worth a comment, much less a moral condemnation. Indeed, in their recommendations of the poem they fail to mention this central element of the poem's allegorical tale entirely.

The epic-length *King Poppy* is in the truest sense a medley. In the words of Bulwer Lytton's biographer Aurelia Brooks Harlan: "Myths of antiquity, history, nursery tale, politics, theories of art, fable, and personal philosophy are mingled, sometimes incongruously. Furthermore, Poppy runs the whole gamut of Lytton's moods, from light-hearted facetiousness to gloomy reflection" (1946, 254). The text also features an idiosyncratic, often ironic prose rubric, which adds an additional layer of abstract meaning. While the poem offers a narrative, though, it is programmatically not a novel in verse, explicitly rejecting realist aesthetics and turning instead to the tradition of folk tales, legends and allegories.

The poem's prologue gives a legendary justification for the poppy's association with death (it was carried with Persephone to the underworld) and with dreams (it was rescued from the underworld by the god of dreams and imagination, Phantasos), but it leaves the poppy in the end exiled on a lonely island and thwarted in its ambition to gain the royal insignia which would mark it king of dreams. The bulk of the work then focuses on the fate of the royal family of the kingdom Diadummiania, in which a mechanical puppet relieves the princess of her royal duties and finally enables her and her father to escape into a perpetual dream under the influence of the poppy. This is often just as whimsically funny as it sounds, but at the same time the text voices some very serious concerns, a mixture that can be rather bewildering. As Herbert Tucker notes in a rare critical consideration of the poem: "*King Poppy* is on one hand a smiling send up of the forms of political sovereignty, on the other hand a solemn defense of the imaginative power that—here already the work's definitive incoherence opens to view—both shapes those institutional forms and erects an inward asylum against all they represent" (2008, 544).

Due to the inconsistency of the poem's allegorical meaning, most reviewers, contemporary and recent, agree in appreciating primarily its satire of Victorian politics, political systems and rational and scientific discourse. Such a focus must lose the poppy from sight. The role that opium plays in the poem, however, is quite intriguing and rather less straightforward than the laudatory conclusion to the poem which I have quoted above may suggest. Mekon (Greek for poppy) is the eponymous protagonist of the poem, but he is relegated mostly to the allegorical frame tale. Moreover, as a flower with little agency, he is a rather doubtful hero. Throughout the poem, his activity is restricted to wheedling and beguiling, and his most heroic characteristic is his persistency and his unfailing belief in his just deserts in the face of derision and opposition. He finally gains the crown and purple mantle he desires, but one is yielded to him by the princess, who feels unfit to face the sorrow of life and the responsibility of ruling her kingdom, preferring to escape into a perpetual (opium-induced) dream, and the other is gained from the lifeblood of a shepherd boy, whose death joins him in dreamland with his love, the princess. The poppy appears as a saviour who promises the princess eternal happiness and relief and who transports the lovesick dying shepherd into a dreamland in which he will "in safety dwell, / Forever loving, and forever loved!" (ll. 4724–25). Nonetheless, the poppy's kingly attributes thus have a sinister origin: "Sorrow gave him one, and Death the other" (ll. 4718). And although the text's representation of the poppy is predominantly and often emphatically positive, it occasionally strikes a more sinister tone:

> Above the purple-cushion'd couch, and close
> To the still maiden's tender ear, pale-lipp'd,
> With writhing neck, in thin green garment clad,
> Like a wan goblin bent beneath the weight
> Of Elfland's gifts (his unawakening hand
> Upon the pillow of its darling child
> Must lay in haste ere morn's return) low crouch'd
> A great white poppy; (ll. 4670–77)

In this scene of seduction, the epic hero has turned into a fairy-tale villain, an evil gnome, with the princess as a version of Sleeping Beauty without the prospect of a prince to come and wake her up. This amounts to an idealised dream of a perpetual gratification of desire, an addict's paradise of a drug high without withdrawal or habituation. Nonetheless, the dreamland of the poppy is a realm without escape, a definitive renouncement of the world and even of life. King Poppy reveals the "Realm of Consolation to a race / Else irremediably miserable" (ll. 509–10), but such consolation is explicitly associated with sorrow and death. The crouching wan goblin suspiciously resembles a parasite and a tyrant, whose "Realm of Consolation" is that of the ultimate addiction in which the drug has replaced both self and world.

Such ambiguity also tinges the strange passage in which the poppy's realm is explicitly named as the seat of poetry. Poetry enters the realm of the poppy as an attribute of the spirit of "Absolute Liberty". This spirit has no place in "[t]he Visible Universe [which] is a system and order of government", since "[t]his Universe existeth only by obedience to the laws that maintain its existence" (Bulwer Lytton 1892, 13). "Absolute Liberty", the source of poetry, however, is a destructive spirit:

> [It is] in the negation of all things that Absolute Liberty seeketh Absolute Truth. [...] And because, for this reason, Absolute Liberty is everywhere repressed, its fierce destructive spirit would fain find in the delirium of intoxication a momentary escape from the tormenting consciousness of an eternal and universal restraint. But there it findeth only Madness, and allied with Madness, it haunteth the Realm of Reality. There man's place is fixed. (Bulwer Lytton 1892, 13)

Hiding from revengeful gods in the poppy, the spirit of Absolute Liberty is transformed “Into that sacerdotal oracle / Of Consolation’s Realm, its listeners call / The ‘Voice of Poësy’” (ll. 257–59):

And “Me”, it sighs, “the world doth still proscribe,
Knowing that I am Freedom, which it fears;
And men mistrust me still, for I reveal
What they deny, refute what they affirm;
And still my law is not as theirs, and still
They deem me lawless, who have here regain’d
My lost legitimate liberty. Beware,
Poor slaves of Reason, who run after Truth!
Wake her not. She is dreaming. Once awaked,
Truth will be folly. Trouble not her dreams!” (ll. 261–70)

This justification of poetry as an assertion of an absolute truth in absolute fancy is as debilitating as it is empowering. Associated with dreams, intoxication, illusion and even madness, the role of poetry is set in explicit and hostile opposition to the “Visible universe”, governed by reason. Poetry is thus ultimately denied an active role in the “Realm of Reality”. Its purview lies in the realm of consolation, which offers an escape from action, responsibility and agency.

King Poppy thus closely associates poetry with opium’s consolatory, dream-inducing and escapist powers, but does so throughout in explicit opposition to a cultural context which values sobriety and reason and which favours science over the imagination, criticising “all the most popular formulas of social and political improvement [whose practical tendency it] is to exclude the imaginative element from the development of character and society, and to ignore its influence” (from a letter quoted in the introductory note to the poem, Bulwer Lytton 1892, vi). The threat of addiction, in this context, while not entirely subdued, is mellowed by the fact that the abdication of self and world it implies is presented in the poem as desirable. Thus, *King Poppy*’s celebration of opium must be understood not only in terms of a “defiantly escapist humanism” (Tucker 2008, 545), or as Lytton’s “only revenge for what he considered his artistic frustration” (Harlan 1946, 262), but also as a deliberate nostalgic reaction against a dominant set of cultural values which goes beyond the merely personal and biographical. Poetry is associated with opium in the poem both on the basis of the dreams it inspires and the consolation it offers. At the same time, the escapist fantasies the allegory indulges in explicitly liberate the dream-world of poetry from a world of practical responsibilities, in which it has no place. In spite of its celebration of this escape, the poem cannot avoid the kind of moral ambiguity that besets a similar scenario in Tennyson's "The Lotos-Eaters". Nor does it entirely manage to subdue the more sombre spectre of drug addiction and death that lurks in the background. At the end of the century, *King Poppy* thus holds on to an alliance between poetry and opium, celebrating their close relationship and their beneficial powers, but it also highlights the moral dubiousness involved in this alliance and bears witness to their exile from an everyday world of practical reality. The proscription the spirit of Absolute Liberty, source of the Voice of Poesy, faces in a world enamoured with order and reason finds its realisation not only in a cultural preference for sober representation over fanciful imaginations, for realist novels over lyrical or allegorical poetry, but also in the increasing regulation of opium distribution and usage.

Beyond the Autonomous Subject

In its positive (albeit to some degree ambiguous) representation of the poppy, *King Poppy* differs strikingly from opium depictions in nineteenth-century prose fictions, which are often steeped in anxieties about the colonial other and the "permeability" of the subject (Margolis 2002, 24) and which increasingly emphasise concerns about addiction. This, I would suggest, is not only due to the Romantic heritage of allying the drug with poetry, but also has its roots in genre logic. In the poems I have discussed here, selves are permeable, no choices are pure and the speakers are always under constraint, both literally and formally. Assumptions about the autonomous individual, which lie at the heart of the aesthetic project of the novel, are not as relevant for the poems. If Tennyson gives voice to his all-engulfing, self-eradicating grief, Thomson imagines the abdication of his separate soul, and Thompson sees himself replaced by his withered dreams, the distinction between interior core and outside control as the basis for the autonomy of the subject is repeatedly abrogated. In *King Poppy*, all human characters lack agency and are directed by the god Phantasos, the ultimate puppeteer. The princess puppet, "Obedient to innumerable wheels / And pulleys in its little bust contain'd" which "rather seem'd / The ideal than the copy of that correct / Assemblage of complete accomplishments / A well-brought-up young lady" (ll. 2553–59), is only the metaphorical culmination of a more widely applicable concept of the individual that is far from autonomous. With its mechanical interior of pulleys and wheels it embodies ambiguities between internal core and external control that continually vex the concept of the individual, in particular in the context of addiction (see Melley 2002, 39). Even the poppy itself becomes the refuge of the spirit of Absolute Liberty, its core inhabited by an alien entity which redefines the poppy's very being. Autonomy and agency are illusory at best for all characters, always excepting Phantasos (i.e. personified imagination) himself, but this lack of agency and permeability of the subject is at no point a matter for concern or anxiety. Since there is little investment in the autonomy of the individual, the potentially addictive tyranny of the drug also appears as less of a threat. At the same time, such a denial of the autonomous self goes hand in hand with an escapist withdrawal from the world and leads to the repeatedly evoked contrast between the poet/dreamer (the poppy) and a productive member of society (the corn).

Opium, Anstie claims, is commonly taken among his contemporaries "by persons who would never think of narcotising themselves, any more than they would of getting drunk; but who simply desire a relief from the pains of fatigue endured by an ill-fed, ill-housed body, and a harassed mind". This practice, he goes on, "must be carefully separated from that kind of narcotic delirium which is sometimes sought for by the literary dilettante". The former is beneficial, "the latter is as truly a poisoning as is alcoholic intoxication" (1864, 149). Such ambiguity in both the drug effect and the moral judgement of its use provides the fertile ground on which the ambiguities of Victorian associations between poetry and opium proliferate. Both poetry and opium offer consolation, freedom and escape, but by association, they may also both wither hopes and numb and paralyse the senses. The relinquishing of the autonomous self and the permeability of the self to external influences, which accounts for much of the anxieties expressed in nineteenth-century prose fiction's depictions of opium, is desired, and often constitutive in the context of poetry. Instead of losing the self in inspired visions, however, the Victorians relinquish the self in oblivion and (suicidal) death. In a cultural climate which builds on notions of individual autonomy and control and values action, such an opiate view of poetry ultimately threatens to devalue poetry's place and function in the world. In a society in which "[t]he disciplined body becomes at times seductively, at times dangerously similar to a constructed object or machine" (Brodie and Redfield 2002, 6), poetry, like opium, at best may serve as a quotidian palliative. At worst, poetry, like opium, threatens to dissolve the individual, to make it, for better or worse, dysfunctional in a functional world.

References

Abrams, Meyer H. 1971. *The Milk of Paradise: The Effect of Opium Visions on the Works of DeQuincey, Crabbe, Francis Thompson, and Coleridge*. New York: Octagon.

Anonymous. 1854. Janus; Lake Sonnets, &c. by David Holt. *The Athenaeum* 1375: 275.

———. 1895. Music in the Public Schools. *The Magazine of Music* 12 (12): 290.

Anstie, Francis Edmund. 1864. *Stimulants and Narcotics, Their Mutual Relations: With Special Researches on the Action of Alcohol, Æther, and Chloroform, on the Viral Organism*. London: Macmillan.

Armstrong, Isobel. 2013. 1832: Tennyson and the Critique of the Poetry of Sensation. In *Tennyson*, ed. Rebecca Stott, 54–75. London and New York: Routledge.

Armstrong, Nancy. 2005. *How Novels Think: The Limits of British Individualism from 1719–1900*. New York: Columbia University Press.

Berridge, Virginia. 1999. *Opium and the People: Opiate Use and Drug Control Policy in Nineteenth and Early Twentieth Century*. Rev. ed. London: Free Association.

Blair, Kristie. 2001. Touching Hearts: Queen Victoria and the Curative Properties of *In Memoriam*. *Tennyson Research Bulletin* 7 (5): 246–54.

Blasing, Mutlu Konuk. 2007. *Lyric Poetry*. Princeton: Princeton University Press.

Bounes, Vincent, et al. 2017. Predicting Morphine Related Side Effects in the ED: An International Cohort Study. *The American Journal of Emergency Medicine* 35 (4): 531–35.

Brodie, Janet Farrell, and Marc Redfield. 2002. Introduction. In *High Anxieties: Cultural Studies in Addiction*, ed. Janet Farrell Brodie and Marc Redfield, 1–15. Berkeley, Los Angeles and London: University of California Press.

Brooks, Peter. 1984. *Reading for the Plot: Design and Intention in Narrative*. Oxford: Clarendon Press.

Budge, Gavin. 2013. *Romanticism, Medicine and the Natural Supernatural: Transcendent Vision and Bodily Spectres, 1789–1852*. Basingstoke: Palgrave.

Bulwer Lytton, Robert. 1892. *King Poppy*. London: Longmans.

Colman, Adam. 2019. *Drugs and the Addiction Aesthetic in Nineteenth-Century Literature*. Cham: Palgrave.

Cunningham, Valentine. 2011. *Victorian Poetry Now: Poets, Poems, Poetics*. Malden and Oxford: Wiley-Blackwell.

De Quincey, Thomas. 2013. *Confessions of an English Opium Eater and Other Writings*, ed. Robert Morrison. Oxford: Oxford University Press.

Freeman, Hannah Cowles. 2012. Opium Use and Romantic Women's Poetry. *South Central Review* 29 (1): 1–20.

Harlan, Aurelia Brooks. 1946. *Owen Meredith: A Critical Biography of Robert, First Earl of Lytton*. New York: Columbia University Press.

Hayter, Alethea. 1968. *Opium and the Romantic Imagination*. Berkeley, Los Angeles and London: University of California Press.

Margolis, Stacey. 2002. Addiction and the Ends of Desire. In *High Anxieties*, ed. Janet Farrell Brodie and Marc Redfield, 19–37. Berkeley, Los Angeles and London: University of California Press.

Martin, Elizabeth. 2015. Opiate. *Oxford Concise Medical Dictionary*. Oxford: Oxford University Press.

Melley, Timothy. 2002. A Terminal Case: William Burroughs and the Logic of Addiction. In *High Anxieties*, ed. Janet Farrell Brodie and Marc Redfield, 38–60. Berkeley, Los Angeles and London: University of California Press.

Milligan, Barry. 1995. *Pleasures and Pain: Opium and the Orient in Nineteenth-Century British Culture*. Charlottesville and London: University Press of Virginia.

O'Gorman, Francis. 2004. *Victorian Poetry: An Annotated Anthology*. Malden and Oxford: Blackwell Publishing.

Platizky, Roger. 2002. 'Like Dull Narcotics, Numbing Pain': Speculations on Tennyson and Opium. *Victorian Poetry* 40 (2): 209–15.

Polsky, Stephanie. 2015. The Novel Ingestion of Opium and Orientalism in the *Mystery of Edwin Drood*. *The Victorian* 3 (1): 1–26.

Roth, Marty. 2002. Victorian Highs: Detection, Drugs and Empire. In *High Anxieties*, ed. Janet Farrell Brodie and Marc Redfield, 85–93. Berkeley, Los Angeles and London: University of California Press.

Sedgwick, Eve Kosofsky. 1993. Epidemics of the Will. *Tendencies*, 130–42. Durham: Duke University Press.

Sivanesan, Eellan, Melvin C. Gitlin, and Keith A. Candiotti. 2016. Opioid-Induced Hallucinations: A Review of the Literature, Pathophysiology, Diagnosis, and Treatment. *Anesthesia and Analgesia* 123 (4): 836–42.

Stevenson, Catherine Barnes. 1982. The Shade of Homer Exorcises the Ghost of De Quincey: Tennyson's 'The Lotos-Eaters'. *Browning Institute Studies* 10: 117–41.

Tucker, Herbert F. 2008. *Epic: Britain's Heroic Muse 1790–1910*. Oxford: Oxford University Press.

Vendler, Helen. 2014. Introduction to *The Art of Shakespeare's Sonnets*. In *The Lyric Theory Reader: A Critical Anthology*, ed. Virginia Jackson and Yopie Prins, 128–40. Baltimore: John Hopkins University Press.

Watt, Ian. 2001. *The Rise of the Novel: Studies in Defoe, Richardson and Fielding*, 2nd ed. Berkeley, Los Angeles and London: University of California Press.

Zieger, Susan. 2008. *Inventing the Addict: Drugs, Race, and Sexuality in Nineteenth-Century British and American Literature*. Amherst: University of Massachusetts Press.

———. 2014. Holmes' Pipe, Tobacco Papers and the Nineteenth-Century Origins of Media Addiction. *Journal of Victorian Culture* 19.1: 24–42.

CHAPTER 4

Dangerous Literary Substances: Discourses of Drugs and Dependence in the Nineteenth-Century Sensation Novel Debates

Sarah Frühwirth

When Wilkie Collins published his best-selling sensation novel *The Woman in White* in 1859, he changed the placid Victorian literary landscape overnight by setting off an avalanche of sensation. Although the public's interest in sensationalism turned out to be fugacious, and also critical interest in the genre was practically non-existent until very recently, it cannot be denied that the nineteenth-century sensation novel was an important part of Victorian literary life or, as Henry Mansel grudgingly admitted, "a great fact in the literature of the day" (1863, 512). Despite the fact that the sensation novel was hailed by the Victorian reading public, the seeming omnipresence of sensational literature

S. Frühwirth (✉)
University of Vienna, Vienna, Austria
e-mail: sarah.fruehwirth@univie.ac.at

N. Roxburgh and J. S. Henke (eds.), *Psychopharmacology in British Literature and Culture, 1780–1900*, Palgrave Studies in Literature, Science and Medicine,
https://doi.org/10.1007/978-3-030-53598-8_4

in the 1860s and 1870s largely met with disapproval by the reviewers, who apprehended a widespread moral decline among Victorian readers. After all, the genre's vivid depiction of social and moral transgressions and its addressing of taboo topics, like murder, adultery and madness, were considered highly inappropriate and deleterious for the supposedly impressionable and vulnerable Victorian novel readers, four fifths of whom were estimated to be women by contemporary reviewers (The Philosophy of 'Sensation' 1862, 345). Therefore, many literary reviewers considered it their duty to voice their misgivings about the emergent genre and to do their utmost to dissuade the Victorian readership from reading sensation novels. Consequently, the seemingly endless mass of sensation novels produced was soon rivalled by an incredible amount of critical articles and reviews published in literary magazines, all of which made it their business to discredit the genre with its readers by pointing out its potentially harmful effects and enumerating its moral and aesthetic defects.

In order to produce an effect with the Victorian readers, the critics of sensation fiction availed themselves of a similar melodramatic, hyperbolic and metaphorical language as the objects of their criticism. Besides employing food and eating metaphors and deploying the language of disease and contagion, many reviewers also reverted to the language of drugs and dependence in their animadversions against sensation fiction. To illustrate the dangers sensation novels allegedly posed to their readers' physical and moral welfare, conservative reviewers frequently refigured the Victorian readers as addicts who are unable to control their desire for sensation (Allan 2013, 90) and likened sensation fiction to psychoactive substances like opium or alcohol. While a number of modern-day scholars have noted "the persistence of this rhetoric of addiction in the 1860s novel debates" (Leckie 1999, 117), only few attempts at getting to the bottom of this phenomenon have been made so far. As Barbara Leckie points out, although "the novelty of the sensation novel [...] certainly demanded a modification of established understandings of narrative reception" (1999, 118–19), it does not follow naturally "that such reception be figured in terms of addiction and disease" (119). According to Leckie, the reviewers' addiction metaphors and similes serve a dual purpose. On the one hand, they can be considered to be "rhetorical gestures that [...] reflect the enormous antipathy to the sensation novel" (1999, 119). On the other hand, their references to drugs and dependence "construct a reader pathologically marked by the effects of disease and addiction" (119), a construction ultimately intended to "consolidate a discourse on censorship" (121).

While this interpretation makes a number of valid points, it does not seem to do entire justice to the complexity of the subject, which is why this chapter is going to take up this issue once again and adduce a number of factors that might have induced nineteenth-century reviewers to equate sensation fiction with potentially dangerous, habit-forming substances and represent its readers as drug users who have lost control over their use. This chapter examines the complex relationship between drugs and sensation fiction and aims to show that psychoactive, habit-forming drugs, like opium or alcohol, were an apt conceptual metaphor for the genre due to their many similarities. As Zoltán Kövecses points out, "[S]imilarity is the basis of metaphor" (2010, 78), which means that "[i]f the two entities [compared] are not similar in some respect, we cannot metaphorically use one to talk about the other" (77). Therefore, when contemporary reviewers compared the sensation novel to drugs and its readers to addicts, they must have had certain "mappings" or likenesses in mind that prompted them to metaphorically connect these two concepts, for instance, the opportunities for mental travel both drugs and sensation fiction afforded their consumers or the threat they allegedly posed to the individual's identity and established boundaries.

Although most of these mappings were probably unconscious, this chapter will argue that the reviewers' frequent recourse to the language of drugs and dependence in their discussions of sensation fiction can also be considered a deliberate strategy to persuade their readers of the baleful nature of sensation fiction. Already in the fourth century BC, Aristotle noted metaphors' ability to "evoke emotions, and influence evaluations" (Sopory and Dillard 2002, 382), which is why "metaphor is ubiquitous in persuasion contexts" (383). By forcing their readers to create a conceptual link between the genre and substances whose pernicious effect on their consumers' physical, mental and moral constitution was a well-established fact in Victorian times, the reviewers tried to open their readers' eyes to the dangers the excessive reading of sensation fiction posed to their mental and physical health as well as their moral integrity. However, it will be argued that the reviewers' employment of drug and addiction metaphors not only "highlighted [their] anxiety with the reading body" (Leckie 1999, 119), but the conceptualisation of sensation fiction as a drug also enabled them to give voice to a wide variety of deep-rooted anxieties concerning social and cultural stability. This chapter proposes the view that because drugs create a dependence on a substance outside the consumer's body and thereby destabilise the boundary between the Self and the Other, they were an ideal means to express the conservative reviewers' fear of the disruptive and subversive forces supposedly inherent in sensational writing.

There are a number of factors that support the conclusion that the reviewers' practice of discursively constructing sensation fiction as a drug not only served stylistic purposes but was in fact part of the reviewers' agenda to sway their readers' opinions about sensation fiction. This, for instance, becomes obvious from the fact that the reviewers' references to drugs do not seem to have been made in a desultory fashion but that they adapted their metaphors and similes methodically depending on which particular aspect of sensation fiction they intended to criticise or which of the genre's harmful properties they tried to expose. Perhaps unsurprisingly, alcohol seems to be the "drug of choice" for many reviewers. After all, although many Victorian medical professionals attached medicinal value to certain types of alcohol, like wine, became increasingly problematised in the nineteenth century. This becomes obvious from the fact that alcoholism was first conceived as a disease in the Victorian age. Thomas Trotter's *An Essay, Medical, Philosophical, and Chemical, on Drunkenness, and Its Effects on the Human Body* (1804) not only constitutes the first systematic treatise on alcoholism but also one of the first to treat habitual drunkenness as an actual disease. In order to illustrate the dangers of habitual alcohol consumption, he detailed the adverse effects of alcohol on the human body and mind and provided a long catalogue of diseases induced by drunkenness: "A train of complaints of the most dangerous nature, at once destroying the body, and depraving the mind, are the certain followers of habitual ebriety. Amidst all the evils of human life, no cause of disease has so wide a range, or so large a share, as the use of spirituous liquors" (Trotter 1804, 139).

While Trotter conceived of alcoholism as an individual problem, habitual drunkenness soon came to be viewed as a social problem as more and more working-class people had recourse to alcoholic drinks because of their sedative effect, or, as Peter Gaskell, a nineteenth-century surgeon, put it: "[T]he labourers, male as well as female, swallow the pernicious draught, and bless it as the boon which relieves them from their harassed sensations" (1833, 116). However, drunkenness, with its impairment of cognitive and motor functions, posed a serious threat to the productivity and efficiency demanded from the industrial workforce, which is why habitual drunkenness soon became one of the most widely commented-on "social evils" in Victorian times.[1] In the light of this fact, it is hardly surprising that reviewers who disapproved of sensation fiction frequently chose alcohol as a point of comparison when pointing out the genre's vitiating properties and the dangers it posed to its readers' mental and physical health. Although a reviewer decried the fact that the "sound old ale" of literary realism, the then accepted mode of writing, had turned to "thin champagne" (Sensation! A Satire 1864b, 86) in the shape of the sensation novel, thereby emphasising the genre's inferiority and "weakness", the majority of conservative reviewers rather tended to stress the genre's high potency when describing its immediate effects on the reader.

After all, sensation fiction did not receive its name just because of its "sensational" subject matters and the "sensational" force with which it hit the nineteenth-century literary market but also due to the physical effects, that is "sensations", it caused in its readers. Besides making their readers' "Flesh Creep" (The Sensation Times 1863b, 193), sensation novels were assumed to "curdle their blood, cause their hair to stand on end, give them 'pins and needles' in the region of the heart, and fix their eyes with a rigid stare for at least twenty-four hours" (New Novels 1864a, 178). In order to impress the detrimental effects of this "unwholesome" mental and physical excitement on his/her readers, the author of an anonymous review in *St James's Magazine* even went so far as to compare the average reader's physical response to sensation fiction, which was supposed to include sweating, an increased breathing rate and blood pressure, trembling and agitation, to a serious medical condition resulting from persistent excessive alcohol abuse or a sudden withdrawal of alcohol, namely delirium tremens:

> We are weary of this unwholesome excitement in any form; but in literature it has become intolerable. We wish it to be understood that we do not object to mental excitement, when it is the natural result of strong intellectual diet. It comes from a wholesome source, and subsides without leaving any ill effects. The sublime creations of genius on an impressionable mind, cannot be appreciated without a sensible elevation of feeling; but this is as different from the feverish sensation sought after by some modern authors, and by their admirers, as is the gladdening exhilaration produced by a glass or two of Moët's purest vintage, compared with the horrible

[1] See essay in this volume by Vanessa Herrmann.

delirium tremens created by a debauch in which the most noxious poisons have entered as elements of intoxication. (The Philosophy of 'Sensation' 1862, 345–46)

Susan Zieger writes, "In the nineteenth century, delirium tremens, with its terrifying hallucinations and tremors, was frequently represented as the dramatic culmination of alcoholism in insanity and death" (2008, 40). At a time when temperance societies and advocates of teetotalism published tracts on the devastating effects of alcohol abuse and temperance novels depicted the dying agonies of alcoholics afflicted with delirium tremens in vivid detail, the analogy between sensation fiction and the effects of long-term alcohol abuse was certainly a well-calculated, powerful image for the purpose of convincing the genre's avid readers of the harmful effect of sensation novels and exhorting them to temper their thirst for sensation. It was certainly no coincidence that reviewers who were hostile to sensationalism likened the genre to distilled spirits rather than weaker types of alcohol, since also various temperance movements tolerated wine and beer while disapproving of distilled spirits. By deploring the fact that the "beverage" of sensation fiction "apparently is not the produce of a vineyard, but of a distillery" (The Philosophy of 'Sensation' 1862, 343), the reviewers assumed the rhetoric of the middle-class "supporters of moderate consumption", who denounced "the common drink of workers while treating their own preferred alcoholic beverages as benign" (Phillips 2014, 197), a practice illustrated by the way in which John MacCulloch, a nineteenth-century physician, pitted spirits and wine against each other: "It is well known to physicians that diseases of the liver are the most common and the most formidable of those produced by the use of spirits. It is equally certain that no such disorders follow the intemperate use of pure wine, however long indulged in" (1817, 151).

Comparing sensation fiction to potables with high alcoholic content rather than to weaker alcoholic beverages served a double purpose. On the one hand, it provided the reviewers with a means to point out the genre's harmful effects in a graphic manner by linking the sensation novel to another, well-established "social evil". On the other hand, it may be argued that equating sensation fiction with distilled spirits, which, as Phillips suggests, were frequently associated with industrialism and industrial workers (2014, 175), constituted a subtle jibe at the genre's roots in a distinctly lower-class genre, namely penny fiction, and its image of being a mass product without any intrinsic value. In fact, many reviewers deplored the sensation novelists' alleged mercantile attitude that made them supply the Victorian reading public with cheap mass commodities of inferior quality for the sake of personal gain:

> No divine influence can be imagined as presiding over the birth of his [the sensation novelist's] work, beyond the market-law of demand and supply; no more immortality is dreamed of for it than for the fashions of the current season. A commercial atmosphere floats around works of this class, redolent of the manufactory and the shop. The public want novels, and novels must be made – so many yards of printed stuff, sensation-pattern, to be ready by the beginning of the season. And if the demands of the novel-reading public were to increase to the amount of a thousand per season, no difficulty would be found in producing a thousand works of the average merit. (Mansel 1863, 483)

Likening sensation fiction to liquor not only served the reviewers' purpose of expressing their concern for the Victorian readers' physical health but also demonstrates their fear for the reading public's mental health and moral purity. Both alcoholism and "dipsomania", a condition characterised by repeated temporary, excessive indulgence in drink, were associated with mental disease. For instance, Trotter not only considered drunkenness "a temporary madness" but also a potential cause of "insanity and idiotism" (1804, 129). A similar view was advanced by Forbes Winslow, a nineteenth-century psychiatrist, who asserted that "more than five-sixths of the insanity in England and Wales is to be ascribed to habitual self-indulgence" (1852, 186). James Miller, a nineteenth-century surgeon involved in the temperance movement, on the other hand, emphasised alcohol's demoralising properties. He believed that "the better part of man sustains both degradation and decay" when indulging in alcohol, until "[u]ltimately, the man becomes a moral idiot" (Miller 1859, 164). Just like alcohol, sensation fiction was believed to have a detrimental effect on its readers' moral and mental faculties. According to popular belief, the genre was "as deficient in moral nutriment as in intellectual pleasure" (The Philosophy of 'Sensation' 1862, 343) and could "neither chasten, refine, nor invigorate the mind" ('Lady Audley' on the Stage 1863a, 244), which is why reviewers also frequently reverted to the language of alcohol when pointing out the genre's mentally and morally debilitating properties. For instance, in an (unintentionally) comic 1867 article in *Tinsley's Magazine*, entitled

"Aunt Anastasia on Modern Novels", "Aunt Anastasia" admonishes her errant, sensation-novel-devouring niece, whose "literary dram-drinking" is perceived to be utterly shameful, to abstain from reading sensation fiction by pointing out the harmful effects this kind of literature supposedly has on the reader's intellect and moral sense:

> The fact is, my dear niece, you are always half-drunk, in the sense of moral intoxication, and that too on the coarsest and most injurious kind of intoxicating literary liquor, the very 'dog's nose'[2] and 'blue ruin'[3] of fiction. There is nothing brilliant, fervent, poetical, in this kind of intoxication; it has the brutalising, enervating effect of dram-drinking; and one of its speediest results is the destruction of the intellectual appetite and digestion. (Aunt Anastasia 2004, 162)

While nineteenth-century reviewers usually employed the language of alcohol when discussing the genre's pernicious effect on their readers' mental health, moral purity and physical well-being, the rhetoric they used when alleging the genre's habit-forming qualities was more varied and also included other types of drugs, like opium, whose habit-forming potential was well established. For instance, Alfred Swaine Taylor, a nineteenth-century toxicologist, remarked that "[t]here is no poison which appears to be so much under the influence of habit as opium" (1859, 592). The concept of addiction, which emerged in the second half of the nineteenth century, not only informed the medical discourses surrounding certain types of drugs but soon also extended to other areas of Victorian life. As Leckie points out, "Reading, especially *addictive* reading, had become a social problem of real concern. The rhetoric of this conclusion obsessively equates reading with stimulants, dram-drinking, opium-eating, and drugs. The reader's appetite for reading grows; the reader wants more and more books" (1999, 112). Like addiction, the readers' craving for sensation novels is imagined as an endless loop of dependence, since the novels for which they hanker do not abate but rather "stimulate the want which they supply" (Mansel 1863, 482–83). A similar point was made by Paget:

[2] A cocktail containing beer and gin.

[3] Low-quality gin; often adulterated and mainly consumed by members of the working class.

> [I]t is come to this, that the many of our day, who make books their stimulants, think themselves at liberty to read anything that amuses them; and they take no count whatever of the time which they so employ. In the absorbing interest of literary dram-drinking, all higher studies are neglected, and the duties of life left unfulfilled. There are not a few young persons who seldom read anything but sensational novels, yet they are reading continually; – that is, in very plain language, they wallow from day to day, amid filth of the most defiling kind. [...] And their appetite grows by what it feeds on, just as the opium-eater requires stronger and stronger doses of the drug that destroys him. (1868, 307–08)

Similar to the user of drugs, who "is imagined to suffer from an uncontrollable desire" (Margolis 2002, 20), the readers of sensation fiction were supposed to be unable to resist their craving for sensation and to need increasing doses of sensational content. W. R. Greg, for instance, described the reading public's appetency for sensation as a "thirst [that] is never slaked" (1860, 405), and Alfred Austin compared the average sensation novel reader to "the drunkard, not only draining his glass, but *licking it out*" (1874, 256). Before the nineteenth century, addiction was a rarely discussed phenomenon that did not generate a lot of anxiety (Berridge and Edwards 1981, xxiv–v). Instead of being labelled as addicts, people with an intemperate craving for intoxicants were conceived as sinners or as lacking the necessary willpower to moderate their desire (Zieger 2008, 4), a conception that also informed the sensation novel debates of the 1860s and 1870s. For instance, in an article in *Temple Bar*, Alfred Austin complained that the three national "bad habits" of "dram-drinking, tea-drinking, and tobacco-smoking" had recently been expanded by a fourth national vice, namely "the vice of reading", a "danger just as real, just as imminent, and [...] yet more deadly, since far more insidious": "Reading, so long a virtue, a grace, an education, and, in its effects, an accomplishment, has become a downright vice, – a vulgar, detrimental habit, like dram-drinking" (1874, 251).

However, this conception of addiction as a bad habit slowly but surely made way for a quite different definition of dependence when "[t]he phenomenon of tolerance was recognized and commented on, along with the allied symptoms of withdrawal" (Berridge and Edwards 1981, 153). By the middle of the century, addiction started to be reconceptualised as an illness, and addicts were increasingly conceived as pathological abnormalities of the general population. For instance, Horace B. Day referred to the opium habit as "the opium disease" (1868, 261) in his treatise *The Opium Habit*, and alcoholism evolved from being conceived as the cause of various diseases to being conceptualised as a "dreadful [...] malady" (Canniff 1872, 19) itself. Since the sensation novel was believed to create a similar compulsion in its readers to the one experienced by users of highly addictive drugs like opium, it is hardly surprising that the genre was also conceived as a virulent infectious disease, a "virus [that] is spreading in all directions, from the penny journal to the shilling magazine, and from the shilling magazine to the thirty shillings volume" (Belles Lettres 1866, 270). According to the anonymous reviewer of an article in the *Westminster Review*, the disease of sensation fiction, similar to other plagues, which usually "occur[] in seasons of dearth and poverty, and attack[] only the poor", breaks out "in times of mental poverty; and afflict[s] only the most poverty-stricken minds" (Belles Lettres 1866, 269–70). It is interesting to note the reviewer's emphasis on poverty that permeates this passage. Like dram drinking or opium-eating, disease was mainly associated with the lower orders of society, which is why the rhetoric of disease and contagion (just like the rhetoric of drugs and addiction) was an ideal discursive strategy to express concerns regarding the subversive elements supposedly inherent in sensational writing. After all, "[f]iction, like contagion, might become the vehicle by which important physical boundaries were breached: distinctions between subject and object, upper and lower bodily strata, upper and lower class, masculine and feminine, food and filth" (Gilbert 1997, 36).

Besides sensation fiction's alleged dependence-creating and tolerance-building properties, there are a number of other mappings between the domain of drugs and the domain of the sensation novel that promoted their consolidation in a conceptual metaphor, for instance, their chief consumers. While inebriety was largely conceived to be a male habit, and temperance tracts written after 1830 focused on urban, working-class males as the chief consumers of alcohol (Phillips 2014, 174), hypodermic morphine consumption was primarily associated with women (Zieger 2008, 127–28; Davenport-Hines 2002, 114). Similar to certain kinds of drugs, also sensation fiction was chiefly associated with the female sex, its production (except for a few notable examples like Wilkie Collins and Charles Reade) and consumption being mainly tied to women (Pykett 1994, 40). Moreover, women are also central to sensation fiction's plots, as many sensation novels feature transgressive, often criminal, female characters who under the cover of a respectable façade threaten to undermine the safety and tranquillity of an upper-class domestic idyll. The genre's exploitation of women's alleged potential for duplicity and disguise constitutes another link between sensation fiction and substance abuse, since addicts, and especially female addicts, were commonly regarded as deceitful.

The "conventional fascination with the deceptive potential of women" (Zieger 2008, 129) that characterised nineteenth-century constructions of the female addict is also palpable in sensation fiction. According to Tara MacDonald, "In sensation novels, the ability to 'step out of one's character' is often figured as a particularly feminine act. That is, sensation fiction frequently suggests that women's identities are more fragmented than those of their male counterparts, and that women are more skilled in the art of disguise and performance" (2013, 128). Although sensation fiction also features male villains who assume a different identity in order to pass unnoticed or trick people, such mechanisms of disguise occur much more frequently with female sensation protagonists. By eradicating her own identity and assuming various other identities, the sensation villainess, usually a beautiful woman from a lower-class background, is able to delude the unassuming members of an idyllic upper- or middle-class household and thereby work her way to the top of the social ladder. Under the guise of being impoverished yet respectable gentlewomen, both Lady Audley, the main protagonist of Mary Elizabeth Braddon's *Lady Audley's Secret* (1862), and Lydia Gwilt, the female protagonist of Wilkie Collins' *Armadale* (1866), are able to marry outside their sphere, and also Isabel Carlyle, the main protagonist of Ellen Wood's *East Lynne* (1861), who is a gentlewoman by birth, is able to return to the family estate after compromising her social standing by running off with another man, a homecoming that would have otherwise been impossible to a social outcast as herself. Interestingly, many characters whose identity is problematised in sensation fiction are associated with drugs and other addictive, psychoactive substances. Lady Audley, for instance, briefly considers committing suicide with a bottle of opium from her medicine chest while Lydia Gwilt takes large doses of laudanum to get some peace of mind and to be able to sleep at night. Also Magdalen Vanstone, the protagonist of Wilkie Collins' novel *No Name* (1862), whose plot is constructed around the main character's loss of identity, thinks about committing suicide with laudanum.

This close connection between drugs and identity is not only a characteristic feature of sensation fiction but is also a central issue in drug fiction and autobiography, in which drugs are usually portrayed as a threat to the user's identity either by multiplying his or her identity or eradicating his or her identity altogether. The addict is frequently depicted as a shadow-like being who inhabits the liminal space between life and death, being too alive to be called dead and too close to death to be considered a fully fledged person. Such ghostlike individuals not only haunt drug fiction and autobiographies; sensation novels are also riddled with spectres who are more dead than alive due to their lack of identity. Lady Audley, who erases her original identity by faking her death and subsequently assumes a number of fake identities, "becomes a living representation of death, a liminal figure hovering between the realm of the living and that of the dead" (Talairach-Vielmas 2013, 28). Similar to Lady Audley, whose lack of identity paradoxically manifests itself in the multitude of fake identities she assumes, also the autobiographical subject of Thomas De Quincey's *Confessions of an English Opium-Eater* (1868) is confronted with a "swirling multiplicity of selves" (Schmitt 2002, 70) in his vivid, oriental dreams: "I was the idol; I was the priest; I was worshipped; I was sacrificed" (118).

Another mapping between drugs and sensation fiction that promoted the genre's metaphorical rendering as a drug is the escapism they both provided their consumers from their monotonous everyday lives. As Richard Davenport-Hines notes, "Drugs are often the recourse of people who are bored or sad or angry" (2002, 13), and taking drugs like opium constituted an easy and cheap way of mental travel to distant times and places. By taking drugs, "the white subject could imaginatively create, enter, and roam Orientalized realms closed to the uninitiated" (Zieger 2008, 34). While De Quincey's exotic dream visions are pervaded by a sense of terror at the foreign scenes ("I thus give the reader some slight abstraction of my oriental dreams, which always filled me with such amazement at the monstrous scenery, that horror seemed absorbed, for a while, in sheer astonishment. Sooner or later came a reflux of feeling that swallowed up the astonishment, and left me, not so much in terror, as in hatred and abomination of what I saw" [1868, 119]), Fitz Hugh Ludlow, the author of the autobiographical novel *The Hasheesh Eater* (1857), paints his mental journeys in a much more favourable light, since they allow him to gratify his "passion for travel [...] with indolence and economy":

> The whole East, from Greece to farthest China, lay within the compass of a township; no outlay was necessary for the journey. For the humble sum of six cents I might purchase an excursion ticket over all the earth; ships and dromedaries, tents and hospices were all contained in a box of Tilden's extract. Hasheesh I called the 'drug of travel,' and I had only to direct my thoughts strongly toward a particular part of the world previously to swallowing my bolus to make my whole fantasia in the strongest possible degree topographical. (64)

Such accounts of drug-induced visions popularised the view that drugs enabled their users to indulge in boundless mental travel to distant places and thereby could offer them an escape from their monotonous everyday lives without leaving the confines of their homes. The same holds true for literature, in particular sensation fiction. According to Zieger, both "reading and opium-eating were techniques of bringing the exotic home" (2008, 19). Although it may be argued that all forms of literature are able to provide their readers with a form of escapism, it cannot be denied that some literary genres are more likely than others to "bring the exotic home", for instance, travel narratives but also other forms of fiction like melodramas or Gothic novels, which were frequently set in foreign countries. While this enabled the reader to experience exotic places and tales of foreign brutality and lust in the safe environment of his or her home (just like drugs enabled their users to mentally travel to distant places), these kinds of fiction, by setting their plots far away in time or in place, at the same time also created an artificial distance or barrier between the reader and the fictional events.

The sensation novel, on the other hand, rather than establishing distance, became famous for its creation of proximity by transferring the melodramatic action from remote, exotic places to an English upper- or middle-class setting. Unlike their antecedents, which set their plots "always at a convenient remove from the identifiable situations of everyday life" (Hughes 1980, 7), the sensation novel, while retaining the romantic/melodramatic mode of its predecessors, almost entirely focused on, as Henry James famously put it, "those most mysterious of mysteries, the mysteries which are at our own doors" (1921, 110), and its "novelty lay in the heroine being, not a picturesque Italian of the fourteenth century, but an English gentlewoman of the current year, familiar with the use of the railway and the telegraph" (112–13). While Gothic novels and melodramas enabled their readers to take mental journeys to exotic places from a safe distance, sensation fiction, like psychoactive substances, truly succeeded in "bringing the exotic home" by setting the sensational events in the reader's sphere of life rather than in Italian castles or oriental harems.

In general, it may be argued that both drugs and sensation fiction promote a subversion of traditional boundaries. Not only do they, as has been argued above, subvert the boundaries between the domestic and the exotic (since both "bring the exotic home"), they can also be said to occasion a destabilisation of the boarders between inside and outside, in-group and out-group. After all, drugs cut across ethnic, gender and social boundaries. As Zieger points out, "We are used to the notion that addiction is a democratic disease, capable of striking anyone, anywhere" (2008, 6). Also De Quincey remarked on drugs' (in this case opium's) democratic properties by claiming that opium brings comfort "to the hearts of poor and rich alike" (1868, 82). The same has been said to hold true for sensation fiction. By maintaining certain key features of middle-class literature, like the concept of poetic justice, but spicing them up with elements of penny fiction (a distinctly lower-class genre), sensation fiction turned out to be "a generic hybrid appealing to servants and masters alike" (Costantini 2013, 99), thereby threatening established class boundaries. Sensation fiction's dangerous potential for undermining traditional social boundaries also manifests itself in certain characteristic plot elements of the genre. In fact, the sensation novel is a very democratic genre since it gives space to members of all social classes (often even giving lower-class characters a voice by letting them narrate a part of the story), allows for class mobility (for instance, the efforts of Walter Hartright, a poor drawing master in Collins' *The Woman in White*, are rewarded with his marriage to Laura Fairlie, a woman far outside his social sphere) and frequently conflates servants with masters (see, for instance, *East Lynne* by Mrs Henry Wood).

The potential blurring of traditional social boundaries and hierarchies was also a common metaphor in the discourses surrounding drugs and addiction. In the nineteenth century, drug habituation was frequently conceived as a form of self-enslavement. By "gradually transforming a self-mastered man into an abject 'slave'" (Zieger 2008, 7), addiction metaphorically undermined the boundary between master and servant. In Victorian times, taking drugs was tantamount to losing control over oneself by creating a dependence on a substance outside one's body.

Applying the rhetoric of addiction to sensation fiction enabled conservative reviewers to express their fear of loss of power and control through a conflation of the Self and the Other: "[T]he anxiety surrounding addiction to opium concerns an insuperable bodily dependence on – and hence conflation with – a presence alien to the self but also, more horribly, alien to the nation" (Schmitt 2002, 71). After all, many of the psychoactive substances consumed by the Victorians were associated with foreignness because they were imported from foreign countries: opium poppies were cultivated in the British colonies in India, tea was imported from China and sugar was produced in the West Indies. The fear of dependence on and contamination by a foreign Other through the consumption of drugs also permeates De Quincey's *Confessions of an English Opium-Eater*:

> The 'English Opium-Eater' announced in the title of *Confessions* is a being constantly subject to disintegration into some Other because of his use of opium; yet the possibility of such contamination is what makes him who he is. The fact that opium is actually imbibed renders this threat particularly visceral: addiction to the drug in this sense amounts to a literal, physical dependence on the foreign. (Schmitt 2002, 83)

The symbolic power shift created by this "dependence on the foreign" was also inherent to the substances themselves. After all, many psychoactive substances that were consumed by the Victorians on a daily basis were not only associated with dominion and supremacy but also carried connotations of revolution, unrest and insurgency: sugar was imported to Britain from the Caribbean, a site of bloody slave revolts; the importation of tea from China was the root cause of the Opium Wars between the two nations; and opium was cultivated in India, which was the site of the 1857 sepoy rebellion, better known as the Indian Mutiny. Moreover, imbibing foreign substances involved a threat of potential degeneration or deracination. As Redfield and Brodie point out, "The figure of the drug addict rested on constructions in the United States and Britain of a foreign other demonically seeking to weaken the nation" (2002, 7), an imagined threat that is also vividly present in sensation fiction.

While a great number of sensation novels feature drugs and drug users in one way or another, there was probably no other author whose novels were as heavily imbued with this motif as those of Wilkie Collins, "a model opium addict" (Foxcroft 2017, 54) who regularly took copious amounts of laudanum himself in order to ease his rheumatic pains.

Although Collins praised the invention of laudanum as a means of relieving his intense physical suffering ("Who was the man who invented laudanum? I thank him from the bottom of my heart [...] Drops, you are darling! If I love nothing else, I love you!" [quoted in Berridge and Edwards 1981, 59–60]), the fear of racial contamination and pollution through drugs (especially opium) also saturates his novels. In fact, many of the characters associated with drugs in Collins' novels either have a foreign background or are marked by a set of foreign characteristics: Count Fosco, the sugar-addicted villain of *The Woman in White*, is an Italian count, and the laudanum addict Lydia Gwilt of *Armadale* spends some months of her youth in service in Madeira, is schooled in France and Belgium and later becomes the travelling companion of a baroness.

Among Collins' extensive oeuvre, *The Moonstone* (1868) is the work that deals the most overtly with drugs and addiction: "If *The Moonstone* is primarily the story of a diamond, its other main subject matter is narcotics. The diamond is bequeathed to his niece by an opium smoker. It is subsequently stolen under the influence of opium. The solution to the crime is staged through the experimental application of opium by an opium addict. And, to cap it all, the novel itself was written by another opium eater" (Sutherland 2008, xx). In constant rheumatic pains and disconsolate at his mother's death, Collins, by his own account, dictated the novel to his secretary under the influence of such heavy doses of laudanum that after the work was finished he reportedly "did not recognise it as [his] own" although he was "pleased and astonished at the finale" (quoted in Foxcroft 2017, 54). Just like in its forerunners *Armadale* and *The Woman in White*, drugs in *The Moonstone* act "as a metaphor personifying the wider general anxieties of pollution, parasitic degeneration, and creeping corruption" (Foxcroft 2017, 58). Nearly all the characters affiliated with drugs are also associated with Otherness and through their Otherness pose a threat to the peace and tranquillity of the Verinder household. Ezra Jennings, a semi-autobiographical character who is a habitual user of opium and has a "mixture of some foreign race in his English blood" (Collins 2008, 367), threatens to expose family secrets through his detective work; the tobacco addiction of Franklin Blake, who is said to have a great number of foreign characteristics due to his extensive journeys abroad, prompts the theft of the precious gem; and John Herncastle, a rumoured opium smoker who spent many years in military service in India, bequeaths the jewel to his unassuming niece, thereby setting in motion the book's sensational plot.

However, psychoactive substances were not only associated with a foreign, exotic Other but also with domestic Otherness, as certain types of drugs carried lower-class connotations. Like dram drinking, which was primarily connected with lower-class men, eating opium began to be conceived as a pernicious lower-class habit:

> The 'problem' of opium was, at various stages in the nineteenth century, seen very much as one of lower-class use [...][.] The belief in working-class 'stimulant' use of opium helped justify the first restriction on the drug in the 1868 Pharmacy Act. [...] The question of who used the drug was central; and the control of lower-class deviance was undoubtedly important. The problem of opium use was in this sense the outcome of the class bias of Victorian society. It was in part a question of social control. (Berridge and Edwards 1981, xxviii)

As this quote from Berridge and Edwards suggests, drug use was not only associated with the individual's loss of control over him- or herself through dependence on a substance but also with the loss of social control. As Costantini points out, the 1860s, which marked the heyday of sensationalism, were a time in which the established English class structure and the unequal distribution of privileges were severely questioned as members of the working class increasingly demanded rights that had hitherto been reserved for members of the middle and upper class, for instance, the right to vote (2013, 100). Although, as Jonathan Loesberg remarks, "the issue of reform makes virtually no appearance in the content of the novels" (1986, 116), sensation novels were nevertheless conceived as "a threat to social stability" (Costantini 2013, 99) due to their fundamentally democratic properties. Thus, it may be argued that by applying the rhetoric of drugs and addiction, two concepts that are equally associated with the lower orders of society and loss of power, to sensation fiction, conservative reviewers tacitly expressed their fear of losing control over the working class through the dissemination of subversive and potentially revolutionary sentiments in the shape of the sensation novel.

This fear of losing power over the working class and the blurring of class boundaries clearly also informed sensation fiction. Besides the obvious link that exists between substance abuse and ethnic and/or racial Otherness, the consumption of drugs in sensation fiction also seems to be closely connected to the consumer's official class status, or rather the lack thereof. Drug use in sensation fiction is usually tied to characters whose class status is of an ambivalent nature. Lady Audley, whose use of laudanum is implied by the fact that she is toying with a bottle of this liquid when contemplating suicide, is a lower-class woman who is able to infiltrate the upper circles of society by means of deception. Sir Percival Glyde, one of the villains in *The Woman in White*, who is a heavy drinker, poses as a baronet, although it later turns out that he is his father's bastard son and does not have any actual claims to the title. Lydia Gwilt, who is an adventuress of the lowest order, takes laudanum on a daily basis. However, not only villainous or criminal lower-class characters are depicted as habitual drug users, also positive characters do not escape this equation. For instance, Gabriel Betteredge, one of the narrators of *The Moonstone*, though being a devoted and loyal servant, suffers from various addictions. Not only does he regularly smoke cigars, also his dependence on his favourite novel, *Robinson Crusoe*, for making decisions marks him as an addict. Although he is not a hardened criminal like Lady Audley, Lydia Gwilt or Sir Percival Glyde, he still poses a threat to the Victorian class system, since he, as steward of the Verinder household, acts as a mediator between the lower- and the upper-class members of the household.

In conclusion, it may be said that the reviewers' recourse to the language of drugs and addiction in their discussions of and animadversions against sensation fiction was not an arbitrary move but a well-calculated discursive strategy. As has been shown, drugs and the sensation novel share a number of similarities or mappings, which is why drugs like opium or alcohol, "social evils" whose pernicious nature was well established, were an ideal conceptual metaphor for the purpose of discrediting sensation fiction with its readers. The rhetoric of drugs and addiction not only enabled reviewers to express a number of concerns for the reading individual's physical and mental health but also more far-reaching anxieties. As this chapter has demonstrated, the rhetoric of drugs and addiction carried associations of gender, class and race, thereby making it a versatile discursive tool to voice a wide variety of concerns related to these social categories. By applying the rhetoric of drugs to sensation fiction and refiguring its readers as addicts, reviewers were able to express fears about the destabilisation of the established boundaries between inside and outside, Self and Other as well as upper and lower class, since both drugs and sensation fiction carried implications of contamination and disorder.

Furthermore, it has been shown that these kinds of anxieties were not confined to the set of conservative reviewers who voiced their misgivings about sensational writing but were obviously also shared by a number of sensation novelists. Although the sale and use of drugs was not criminalised in Victorian times, it is significant that mostly foreigners or lower-class characters who attempt to intrude into the lives of peaceful upper-class families and use fake identities to move freely between social spheres are depicted as drug users. Despite the fact that some sensationalists (most notably Wilkie Collins) were habitual drug users themselves, the users of drugs in their works are almost always suspect figures who pose a threat to the established social order. This, on the one hand, shows how deeply rooted such fears were in the Victorian collective consciousness and, on the other hand, how successful the conservative reviewers' discursive attacks were. Indeed, their efforts to discredit the genre and halt the disruptive forces supposedly inherent in sensational writing were seemingly crowned with success when interest in the genre began to flag a few years after its heyday. However, as a number of scholars have noted, residues of the sensation novel remained, giving rise to new forms of fiction, such as New Woman or detective fiction, in whom the democratic tendencies of the genre lived on, thus serving as a forceful reminder that demands for equality cannot be contained by elaborate discourse or reasoning but will always find an outlet.

References

Allan, Janice M. 2013. The Contemporary Response to Sensation Fiction. In *The Cambridge Companion to Sensation Fiction*, ed. Andrew Mangham, 85–98. New York: Cambridge University Press.

Anonymous. 1862. The Philosophy of 'Sensation'. *St James's Magazine* 5: 340–46.

———. 1863a. 'Lady Audley' on the Stage. *The London Review* 6: 244–45.

———. 1863b. The Sensation Times. *Punch, or The London Charivari* 44: 193.

———. 1864a. New Novels. *The Athenaeum* 1919 (2): 177–78.

———. 1864b. Sensation! A Satire. *Dublin University Magazine* 63: 86–89.

———. 1866. Belles Lettres. *The Westminster Review* 86: 268–80.

———. 2004. Aunt Anastasia on Modern Novels. *Tinsley's Magazine* 1 (1867): 308–16. Rpt. in *Sensationalism and the Sensation Debate*, ed. Andrew Maunder, 160–70. London: Pickering and Chatto.

Austin, Alfred. 1874. The Vice of Reading. *Temple Bar* 42: 251–57.

Berridge, Virginia, and Griffith Edwards. 1981. *Opium and the People: Opiate Use in Nineteenth-Century England*. London: Penguin Books.

Canniff, William. 1872. *The Effects of Alcohol Upon the Human System: An Essay Upon the Cause, Nature and Treatment of Alcoholism*. Toronto: Adam, Stevenson and Co.

Collins, Wilkie. 2008. *The Moonstone*. Oxford: Oxford University Press.

Costantini, Mariaconcetta. 2013. Sensation, Class and the Rising Professionals. In *The Cambridge Companion to Sensation Fiction*, ed. Andrew Mangham, 99–112. New York: Cambridge University Press.

Davenport-Hines, Richard. 2002. *The Pursuit of Oblivion: A Global History of Narcotics*. New York: W. W. Norton.

Day, Horace B. 1868. *The Opium Habit, With Suggestions as to the Remedy*. New York: Harper and Brothers.

De Quincey, Thomas. 1868. *Confessions of an English Opium-Eater, Being an Extract From the Life of a Scholar*. Boston: Ticknor and Fields.

Foxcroft, Louise. 2017. *The Making of Addiction: The 'Use and Abuse' of Opium in Nineteenth-Century Britain*. London: Routledge.

Gaskell, Peter. 1833. *The Manufacturing Population of England, Its Moral, Social, and Physical Conditions, and the Changes Which Have Arisen From the Use of Steam Machinery*. London: Baldwin and Cradock.

Gilbert, Pamela K. 1997. *Disease, Desire, and the Body in Victorian Women's Popular Novels*. Cambridge: Cambridge University Press.

Greg, W. R. 1860. French Fiction: The Lowest Deep. *The National Review* 11: 400–27.

Hughes, Winifred. 1980. *The Maniac in the Cellar: Sensation Novels of the 1860s*. New Jersey: Princeton University Press.

James, Henry. 1921. *Notes and Reviews: A Series of Twenty-Five Papers Hitherto Unpublished*, ed. Pierre de Chaignon la Rose. Cambridge, MA: Dunster House.

Kövecses, Zoltán. 2010. *Metaphor: A Practical Introduction*, 2nd ed. Oxford: Oxford University Press.

Leckie, Barbara. 1999. *Culture and Adultery: The Novel, the Newspaper, and the Law, 1857–1914*. Philadelphia: University of Pennsylvania Press.

Loesberg, Jonathan. 1986. The Ideology of Narrative Form in Sensation Fiction. *Representations* 13: 115–38.

Ludlow, Fitz Hugh. 1857. *The Hasheesh Eater: Being Passages From the Life of a Pythagorean*. New York: Harper and Brothers.

MacCulloch, John. 1817. *Remarks on the Art of Making Wine, With Suggestions for the Application of Its Principles to the Improvement of Domestic Wines*, 2nd ed. London: Longman, Hurst, Orme, and Brown.

MacDonald, Tara. 2013. Sensation Fiction, Gender and Identity. In *The Cambridge Companion to Sensation Fiction*, ed. Andrew Mangham, 127–40. New York: Cambridge University Press.

Mansel, Henry. 1863. Sensation Novels. *The Quarterly Review* 113: 481–514.

Margolis, Stacey. 2002. Addiction and the Ends of Desire. In *High Anxieties: Cultural Studies in Addiction*, ed. Janet Farrell Brodie and Marc Redfield, 19–37. Berkeley, Los Angeles and London: University of California Press.

Miller, James. 1859. *Alcohol: Its Place and Power*. Philadelphia: Lindsay and Blakiston.

Paget, Francis Edward. 1868. Afterword. In *Lucretia; or, The Heroine of the Nineteenth Century: A Correspondence, Sensational and Sentimental*, 297–310. London: Joseph Masters.

Phillips, Rod. 2014. *Alcohol: A History*. Chapel Hill: University of North Carolina Press.

Pykett, Lyn. 1994. *The Sensation Novel from 'The Woman in White' to 'The Moonstone'*. Plymouth: Northcote House.

Redfield, Marc, and Janet Farrell Brodie. 2002. Introduction. In *High Anxieties: Cultural Studies in Addiction*, ed. Janet Farrell Brodie and Marc Redfield, 1–15. Berkeley, Los Angeles and London: University of California Press.

Schmitt, Cannon. 2002. Narrating National Addictions: De Quincey, Opium, and Tea. In *High Anxieties: Cultural Studies in Addiction*, ed. Janet Farrell Brodie and Marc Redfield, 63–84. Berkeley, Los Angeles and London: University of California Press.

Sopory, Pradeep, and James Price Dillard. 2002. The Persuasive Effects of Metaphor: A Meta-Analysis. *Human Communication Research* 28 (3): 382–419.

Sutherland, John. 2008. Introduction. In *The Moonstone*, by Wilkie Collins, vii–xxix. Oxford: Oxford University Press.

Talairach-Vielmas, Laurence. 2013. Sensation Fiction and the Gothic. In *The Cambridge Companion to Sensation Fiction*, ed. Andrew Mangham, 21–33. New York: Cambridge University Press.

Taylor, Alfred Swaine. 1859. *On Poisons in Relation to Medical Jurisprudence and Medicine*, 2nd ed. London: John Churchill.

Trotter, Thomas. 1804. *An Essay, Medical, Philosophical, and Chemical, on Drunkenness, and Its Effects on the Human Body*, 2nd ed. London: Longman, Hurst, Rees, and Orme.

Winslow, Forbes. 1852. Crime, Education, and Insanity. *The Journal of Psychological Medicine and Mental Pathology* 5 (18): 161–208.

Zieger, Susan. 2008. *Inventing the Addict: Drugs, Race, and Sexuality in Nineteenth-Century British and American Literature*. Amherst: University of Massachusetts Press.

PART II

Rethinking the Pharmacological Body: Drugs and the Borders of the Human

CHAPTER 5

Blurring Plant and Human Boundaries: Erasmus Darwin's *The Loves of the Plants*

C. A. Vaughn Cross

This essay discusses the pivotal role of Erasmus Darwin's first scientific poem *The Loves of the Plants* (1789, 1791–92) in shaping psychopharmacological ideas—known drug effects—at the end of the long eighteenth century.[1] Deploying poetry and prose in a "conversational arc between modern science and literature" (8), the heuristic amusée contributed to early modern readers' scientific and political ideas about drugs, pain and human health. More than a game to benefit a singular reader, however, reading *LOTP* "under the Banner of Science" (*LOTP* Frontispiece) about interaction with "toxics" (*LOTP* 1. 196–97) should enhance meaningful connections of as many beings as possible. More than a simple taxonomy of "good" or "bad" drugs sought by later Romantics and Victorians,

[1]Although each of *LOTP*'s four cantos is numbered as I followed by line numbers, they are distinguished here as I, II, III or IV.

C. A. Vaughn Cross (✉)
Department of Geography and Sociology, Samford University, Birmingham, AL, USA
e-mail: cavaughn@samford.edu

N. Roxburgh and J. S. Henke (eds.), *Psychopharmacology in British Literature and Culture, 1780–1900*, Palgrave Studies in Literature, Science and Medicine,
https://doi.org/10.1007/978-3-030-53598-8_5

LOTP presented the "*pharmakon*-like indeterminacy" (Boon 2002, 24) of ninety-plus botanicals able to alter human perception and behaviour. From single fungi to admixtures of "affecting" liqueurs, gasses and powders containing caffeine, psilocybin, opium and/or alcohol, *LOTP* depicted multiple points of view within systems of drugs and the human consumption of them.

This essay considers why *LOTP* stood out among contemporary literary efforts to advance general knowledge of drug efficacy prior to nineteenth-century sumptuary laws that regulated drug processing, prescribing and consuming (DerMardersian 1995, 19). It demonstrates how, by conjugating the forces of Hermes (medicine) and Aphrodite (love/compassion), Darwin attempted to disseminate biotic knowledge as a democratic "social force" (Kuhn 1971, 283) and civic good, a theme which continued in his later works, *Phytologia*, *Zoonomia* and *The Temple of Nature*, and initiated his influence on Romantic perceptions of drugs. The essay concludes that *LOTP's* representation of interspecies intoxication and drugs as cause of or remedy for isolation advocates a moral economy of civic concern for as many beings as possible rather than for a colonial society of self-centred or apathetic individual drug consumers.

Garlands of Knowledge

The early modern psychoactive revolution was one of "the signal events in world history", as Westerners sought "progressively more [...] potent [...] means of altering their ordinary waking consciousness" (Courtwright 2001, 2; 6). Concomitant technological, economic and socio-political revolutions reified humans' insatiable appetites and capacities to pursue new energies. By the end of the eighteenth century, recreational and medical experiments with psychostimulants blurred. Galenic drug preparations remained inconsistent. Scientific theories about drug habituation and dependency were evolving (Maehle 1999, 184). Most quality control of psychostimulants was maintained by guilds (Spary 2012, 151–53). Because "toxics" might soften human "chains of conquest with the wreaths of love" or erupt like "thunder" (*LOTP* 1. 196–97), knowledge of how to elicit and manage these effects became a form of currency. Medical practitioners were not the only ones trafficking in such knowledge. Eighteenth-century farmers, cooks, distillers and apothecaries exchanged new drug knowledge and technologies as entrepreneurial opportunities expanded for botanist-merchants, grocer-confectioners, herbal authors and publishers. Financial profit-motives often determined which botanicals were studied and promoted as therapeutic and/or recreational (Kuhn 1971, 280–86).

As ideas fluctuated about who could or should manage new drugs and accoutrements in the emerging "medical machine" (DerMardersian 1995, 18–19), the polymathic physician Erasmus Darwin (1731–1802) harnessed and systemised encyclopaedic pharmacognosy for general readers in didactic poetry and prose as a civic good. Viewing himself as a link in a chain of scientific knowledge across cultures and centuries (*LOTP* Inter. II), Darwin sought to increase modern citizens' understanding of drugs' (un-)predictability while bolstering hope and confidence for managing them. Derived from the Greek terms *pharmakon* ("drug") and *gignosco* ("to acquire knowledge of"), *pharmacognosy* refers to the knowledge of plants, animals and minerals as sources of medicinal drugs. French scientist Denis Dodart connected elements to *material medica*'s therapeutic efficacy in 1676, but the term *pharmacognosy* was not commonly used until early nineteenth-century Continental publications *Lehrbuch der Materia Medica* (1811) and *Analecta Pharmacognista* (1815). First published with popular success in 1789 and expanded as part of *The Botanic Garden* (1791–92), *LOTP* contributed to this developing "science of drugs [...] studied to help other living beings" (De Pasquale 1984, 9–16) by increasing human cognisance of the interactive effects of plants' bio-chemistry, environmental and ethnographical contexts, animal physiology and human socio-economic systems.

Like William Buchan's 1769 declaration in *Domestic Medicine* that individuals "may be their own physicians" (Grinnell 2006, 231), Darwin assumed readers were subjects of their own health experiments. Generally, an advocate of self-determination politically, Darwin emphasised people's need for scientific knowledge about alkaloids in an age of libertine self-experimentation, proliferating quackery and predatory charlatanism. Critical of frameworks that he believed preyed on people's vulnerability and gullibility, such as religious superstitions, hucksterism and those who "believed themselves to be affected by an invisible agent, which the Enchanter called Animal Magnetism!" (*LOTP* III. 7), Darwin addressed botanical knowledge to the "body politic" (Grinnell 2006, 224) as *LOTP* affirmed or undermined their traditional herbals, folk medicine and fads. This made Darwin one of the most influential English writers combining "modern science and technology in verse" (Robinson 1954,

314). With *LOTP*, Darwin became the most explicit natural philosopher of his generation writing on popular poetic herbals (Mahood 2008, 94–96) to cultivate a "promiscuity of knowing" (Green 2007, 205) about what people ate, drank, inhaled and injected (Barker-Benfield 1992, 26). Engaging the reading public about potential benefits and dangers of human-plant interactions amid rapidly shifting "social preferences and perceptions of efficacy" (Grinnell 2006, 230), circulating pharmacognosy among various "domains of epistemological possession" by specialists and non-scientists alike (Spary 2012, 192) was a democratic move.

LOTP simultaneously inculcates a view of readers as participants in a network of botany-hunters, farmers, brewer/distillers, merchants and apothecaries as it cultivates trust in physician-scientists' expertise to assess drug doses' risk-benefit ratios and overall value to wellbeing. By looking for clues in *LOTP*'s textual jigsaw of ninety-six vignettes of images and associations and footnotes of information strung together like a textual daisy-chain across four cantos intersected by three reflexive "interludes" about them, readers could discover Dr Darwin's explicit and implicit prescriptions and warnings involving mood-altering seeds, leaves, oils, nuts and gasses. Given the variability and volatility of eighteenth-century experiments with drugs, *LOTP* purposefully demonstrates the value of scientific taxonomies. "Divine HYGEIA!" reappears throughout *LOTP* as the Enlightenment physician-apothecary familiar with numerous ways of sorting and combining plants' pharmacological virtues according to Galen, Paracelsus, Bacon, Linnaeus, de Jussieu and other "new rules" for classifying effects of botanicals already known and newly encountered among Europeans (De Pasquale 1984, 7; DerMardersian 1995, 16). *LOTP* braids these old and new paradigms with Darwin's own medical theories and Jacobin politics. Its most obvious paradigm, Linnaeus' sexual taxonomy of plants, orders the poem's vignettes in increasing numbers of stamens/males and pistils/females; but Francophile Darwin entwined Linnaeus' system in the French "naturalism" of Bernard and Antoine-Laurent de Jussieu, grouping plants by their flowers and fruits as well as their pistils and stamens (Browne 1989, 598–99; Johnson 2018, 42–44) to allow more room for more recipes and social commentary.

Interspecies Inebriation

In *Eating the Enlightenment*, E.C. Spary observes that a significant "problem in the study of material substances" such as alcohol, opium, caffeine or psilocybin is reconciling "the presumed identity of a given substance in space and time" with its "variety of appropriations of that substance in different cultural settings" (Spary 2012, 192). To avoid anachronistic "commodity biography" or narratives that assume a drug's "continuous and uninterrupted historical identity" (Spary 2012, 192), it is useful to conceptualise "toxics" as Darwin did, as quasi-living substances and actants with adaptive, transforming powers and points of view in animal/human lives and ecological systems. It is also worth noting terms Darwin did and did not use for drugs in *LOTP*.

The word *drug* appears only once in *LOTP*, in reference to malaria treatments (III. 349), and the term *pharma* never appears. Forms of the word *stimulant* appear four times, three of which are about salt. Versions of *health* and *cures* appear only a few times each. Darwin refers to *opium* iatrically five times and *narcotic* qualities twice. He plays poetically with *ether* more than twenty times. In each of the ten instances where Darwin uses the term *toxic* as either a noun or adjective, it is in a form of *intoxicating*. The word *poison* appears at least fifty times with reference to a variety of botanical substances. As F.W.J. Schelling would write in *First Projection of a System of Nature Philosophy* (1799), no poison exists "in itself" (Grinnell 2006, 241), only in relation to its effect on another. Notably, the psychoactive drug Darwin considered the greatest "poison" to personal and public health—alcohol—never appears in *LOTP* as the term *alcohol*. Yet through prose and poetic images, *LOTP* vividly portrays the "spirit" and powers (Maehle 1999, 134) of various fermented and distilled substances compared and contrasted to other drugs.

By linking botanical, ethnographic and socio-economic factors involved in the development of alcohol, opium, caffeine and other psychoactive substances into commodities, *LOTP* demonstrates their "overlapping frameworks for classifying and ascribing value" (Spary 2012, 192). Retaining their power to effect pain and pleasure in other beings after being processed by them, psychoactive substances maintain their own point of view even as they inhabit "more than one social world of meaning and use at once" (Spary 2012, 192). *LOTP* teems with intoxicated fish, birds, beetles and humans "inebriate with their zeal" after eating stimulating turmeric (I. 65–68), muschus (II. 229–66) or coral moss (IV.354–60). It explores why and to what effects other creatures, especially humans, return to consume plants' defensive alkaloid essences and innovate ways to extend such intoxicating experiences through teas, jellies, syrups and powders. Every sentient being is in a constant state of transmutation, potentially altering others' existence as well (Valsania

2004, 349) until the intoxicant and intoxicated who consume it become indistinguishable from each other.

Darwin blurs human-plant boundaries on *LOTP's* first page, reversing Ovid's *Metamorphoses* in which humans translate into plants (as physical bodies eventually do in decomposition). Declaring that whereas Ovid "did by art poetic transmute Men, Women [...] into Trees and Flowers", Darwin has "undertaken by similar art to restore some of them to their original animality" (Frontispiece). His plant personifications before and after their translation into commodities and consumption by other species elucidate the "complex phytochemistry of biologically active constituents" of drugs (DerMardersian 1995, 18). Demonstrating botanical substances' sensitivity, (ir-)rational movement, organisational behaviour, sexual agency and "vegetable storgé" for other creatures (*LOTP* I. 61), *LOTP* suggests that drugs come from "thinking, active, and intelligent creatures" with "emotions", as later scientists have asserted (McKertich and Shilpa 2016, 36–37; 41). Ingesting such living substances extends their energy and power in those who "gambol" with them, and we metabolically become what we eat (and drink, inhale and inject), as *LOTP*'s finale vividly reiterates.

Measures of Individual and Civic Pleasure and Profit

LOTP's "trip" to "ambrosial" (I. 327, 85) health begins with a kitchen herbal of mildly narcotic stimulants, including amphetamine-like turmeric and traditional mood-enhancing *Melissa* (1. 5–68) as "Divine HYGEIA!", and then she rhythmically alternates her apothecary "wand" between stimulants and sedatives. After a wild "ride" with mushrooms, Canto II features opium and gasses in pastoral, picturesque reveries interrupted by tea before Canto III's gothic nightmares of compulsions and convulsions follow. Offering hope that better scientifically managed compounds may edify individual and civic wellbeing, Canto IV resolves psychoactive horrors and thrills into a corporate sublime. Such overall wellbeing is like a whole synaesthesia of "coloured juices" that "taste" like echoes of "Æolian strings" and look and sound like "reflective walls" (IV. 279–98), serving as many people as possible and not just a singular reader's pleasure.

To increase readers' knowledge of drugs' dangers and therapeutic potential, Darwin pedagogically connected information on plants to neo-Classical, Romantic and popular poetic images. Similar to William Curtis' botanical pedagogy in *Flora Londinensis* published in the 1770s and 1780s, *LOTP* even more extensively uses literary associations, resemblances and metaphors from "our subjective archive of sensory experience" (Dugan 2011, 4–5) to stimulate readers' imagination, memory and understanding. *LOTP* weaves puns and other associations of plants' Latin names, common names, morphology and effects. Reading about *Collinisinia*'s pre-commodification erotic pleasures and mint extracts could be enjoyable and physically profitable (I. 51–56). At the same time, botanist-merchants like Peter Collinson who sold these elixirs profited financially from product placement in *LOTP*, fulfilling one of the many *paronomasia* (multiple puns) with which *LOTP* begins: the "rule of Horace" to mix pleasure and profit, whether in reading, in medicine or in other ways.

Despite Linnaeus' rejection of the Doctrine of Signatures and other theories of resemblance as "false and delusive", Darwin employed what Michel Foucault calls the "constructive role" of plant-human resemblance as a beneficial pedagogical tool to engage readers in a game of connections (Johnson 2018, 43; Foucault 1994, 17). *LOTP*'s botanic terms and images wind around themselves and other species like hemp threads. Indeed, the second edition's associations of cannabis' morphology and virtues depict it spiralling like a textile, unspooling a "slow [...] breezy strand" with "streams dishevell'd in her hand" (*LOTP* IV. 111–12). Potential psychoactive effects are merely implied by how one chooses to read Darwin's description of the cannabis plant/spinner/ingester as one-in-the-same: turning back and forth in "airy sweeps" as it "inchants" with a "sweet assuasive smile", a "fringed lid" and a "warm cheek" that "glows" and "burns" (IV. 114, 79–132).

Similarly, Tremella fungi's jelly-like "quivering"—alone or when ingested by frogs, birds and humans—is indistinguishable from its psychedelic effects: "No dim electric streams" lasting "12 to 16 hours" and why mushroom-eating Siberians exchange their hallucinogenic urine as currency (I. 427–84). The intoxicant/intoxicated "flies [...] stops [...] pants [...] looks behind" (I. 445–52). Suggesting Romantics' intrigue with the "empowered-because-disempowered imagination" (Priestman 2013, 129), mushroom-eaters may hear "Tritons sound their twisted shells" (I. 414–26) and exclaim "I sink, I fall!" though physically still. Other substances experience and effect "diamond eyes" and "ever-changing dyes" (IV. 215–16). Like the lotus plants "trembling on the tide" and changing appearances, the edible rush ingester might try "a thousand forms" of identities (IV. 443–67), which is one of many reasons to consider Darwin the source for the Wordsworthian emphasis on "interior body consciousness" (Budge 2007, 280; 296).

In "perhaps the first [period] in European history when a link was widely made between artistic and intellectual creativity of all kinds and the working of psychostimulants" (Vickers 2015, 328; 335), *LOTP* declares experimentation with edible rush and similar "plants of passage" to be "well worth further enquiry" (IV. 200). Changing initially disorienting perspectives effected by plants' alkaloid defence against those consuming them could open "new forms of spatial awareness" (Taylor 2011, 1) and perhaps generate innovative, useful ideas of scientists, inventors, artists and politicians for the civic good. One of *LOTP*'s many Jacobin-flavoured passages is like Thomas Paine's earlier philosophical-poem of couplets, under the *nom de plume* "Snow Drop", which appears in *LOTP*, about change and revolution. It analogises "Winter" to one who inevitably "leaves his empire to the Spring" of liberalism, stimulated by the "leaves" of revolutionary writers' pages and whatever might be in their coral smoking pipes (Paine 1870; 1748 French *Complete History of Drugs*; *LOTP*, IV. 350). *LOTP* offers multiple puns on the literary "passage" associated with sea travel and inner psychoactive journeys on the "brink" (IV. 299–320) of liminal consciousness. Adventurous "gambolers" might "plunge [...] beneath the main" (IV. 210) of ordinary perception into an "unfathom'd deep" (II. 295–98) and unlock access to the "inmost bowers" (I. 470) of enlightenment.

LOTP demonstrates Darwin's own experimentations with affecting gasses in addition to the nitrous oxide circulating among Davy, Beddoes, Coleridge, Southey and Barbauld. Multiple passages on gasses and gasping, and nods towards Lavoisier and other French scientists who linked olfaction, oxygène theories and public health (Dugan 2011, 186) reflect Darwin's role in the developing pneumatic medicine, which infused revolutionaries' "air of reform" and Romantics' "atmosphere" of hope and angst (Grinnell 2006, 229). *LOTP*'s final philosophical interlude is illumined with glass chimney Quinquets of French pharmacists Antoine Quinquet and Aimé Argand, which allowed better inhalation of psychoactive smoke (Inter. III). *LOTP* opines on the rainbow spectra of blooming algae's changing ethanol complexes and likens breathing these gasses to Pan blowing "changeful rings" from "his sevenfold reed" with a "glowing lip" (I. 320, 345–46; II.328). Hues begin "[c]losed in a green recess" of pastoral grasses and then fragrantly rise as "blue smoak" from the chemist's chimney (IV. 85–92) to float across the "ethereal plain" (l. 226, 339, IV.495). Inhalers feel like "plumy sails" and "fanning gales" (IV. 455–66) ready to "sing" to "amorous Echo" (IV. 189–98). The "floating lover" (IV. 455–66) may be especially fond of the drugs and druggists who make these sensations possible, perhaps anticipating Wordsworth's "vital breeze" that "is become/A tempest, a redundant energy/Vexing its own creation" (*Prelude*, Budge 2007, 296).

Operating in the cult of sensibility that considered excitability, or irritation, the "master principle in nature" (Vickers 2015, 327), *LOTP* reflects Darwin's obsession with distinguishing paroxysms as signals of pain, pleasure or both (i.e. the sublime). A popular health paradigm among Darwin's network of Edinburgh-trained physicians was Brunonianism, which looked for over- or under-stimulated, blocked or depleted nervous energy (Grinnell 2006, 237; Vickers 2015, 330). Darwin was familiar with *Elementa Medicinae* (1780) by John Brown, a laudanum addict who died the year before *LOTP* was published. Darwin protégé Thomas Beddoes later translated *Elementa* from Latin into English around the time Darwin published *Zoonomia* (1794–96). Darwin then asserted that while he considered *Elementa* "a work (with some exceptions) of great genius", any "coincidence" between his work and Brown's should "be considered as [...] arrived at by different trains of reasoning" (*Zoonomia* XII. Part II; Lawrence 1988, 11; Barfoot 1988, 25).

Brown's theories may have been simpler than Darwin's (Page-Jones 2018, 90), but Brunonian ideas about energy continued to influence writers influenced by Darwin (see Budge 2007; Vickers 2015; Grinnell 2006). In a "century devoted to the intellectualisation of sex" and "erotomania", *LOTP* exemplified the late-eighteenth-century *au currant* search for stimulants of the pleasure principle to open a "wider libidinal lens" of creativity for social progress (Porter 1982, 5–6; Taylor 2011, 9). "Nothing was more important to the life and work of Erasmus Darwin than erotic love" (Komisaruk 2014, 1), and *LOTP* uses erotic images pedagogically to emphasise drugs' potential as aphrodisiacs and aphrodisiacs' potential to improve health. Hygeia's "potent wand" of pen, pipe and pestle conjures happiness elixirs of "silver rain", white powdered "candied stems" and "ice-flower gems" (IV. 85–92; 217; 34–36; 235–78). Textual boughs form "mystic knots of love" compounds that one could try at home or request from a chemist: "affecting" blends o f Proud Pink, honey of arum and clove, which Darwin noted possess a "narcotic quality" (IV. 279–98; 299–320). In *LOTP*, ingesting rejuvenating blends of marigold (calendula), orange lily, Indian Pink and nasturtiums containing psilocin could simulate holding electric eels (IV. 43–78; 136 footnote). Other Romantic recipes offer mildly psychedelic effects of eugenol and oil extracts of cactus grandiflorus, Arabian jasmine, geranium triste and hibiscus (III. 184; IV. 13–42; 299–320).

Public or Private Opiate Reveries

Perhaps *LOTP*'s most famous drug passage is that of the poppy (II. 267–304). Darwin placed the "queen" of sedatives ("Sleep and Silence" II. 268–69) in Canto II surrounded by plants from *The Horologue (Watch of Flora)* that emphasise suspensions or distortions of time (II. 165–98) even as Canto II "measures" drug doses as analogous to clocks, music, poetry and electricity. *LOTP* describes opium metabolisers like those who gambol with mushrooms or affective gasses: "Froze by inchantment" of "forms of Fancy and Dreams" that pass inwardly "in scintillating streams", ingesters experience an opened capacity to hear "viewless lutes" and "aërial voices" (II. 271–75).

By 1789, a genre of "opiology literature" had developed in medical writing on the Continent (Maehle 1999, 129), as opium had become an "effective analgesic" and alternative to such dangerous practices as bloodletting (Vickers 2015, 331). It was commercially popular in laudanum tinctures (alcohol) and "wellbeing" powders and confections sold with other experimental ingredients for flavour. Darwin's repetition of these passages in *Zoonomia* and his liberal prescriptions in his medical practice formed a "spider's web of opium" that "cast its viscid reticulation from point to point" around a social network including daily ingester Wilbur Wilberforce, Baptist preacher Robert Hall, Samuel Taylor Coleridge, Darwin's in-law Josiah Wedgwood and his son Tom, and Darwin's first wife, Mary (Boon 2002, 90–91).

LOTP eroticises procurement of the "finest opium" from the East as humans catching poppy's pearly "drops" (LOTP II. 270) of treasure in shells, one *LOTP*'s twenty images of Aphrodite's shell also found in Darwin's self-designed family crest. *LOTP* further portrays opium's positive iatric powers as Hermes' caduceus bringing sufferers out "[f]rom the dark regions of the imprison'd dead" (II. 291–92). Anecdotes of

Indian messengers and horses travelling hundreds of miles on nothing but rationed bits of opium doubtless piqued the interest of English readers looking for ways to extend their productivity: "In small quantities [opium] exhilarates the mind, raises the passions, and invigorates the body" (II. 270). As De Quincey later observed in *Confessions of an English Opium-Eater* (1821), by freeing the mind and body from pain, opium "liberates the author" (Fulford 1999, 237) to pursue other activities.[2] Significantly, like poppy's reproductive morphology, many males and females could enjoy opium's effects together on "celestial beds" like gardens or couches (II. 271–75); medically and civically, opium should benefit as many people as possible as a social good by connecting rather than isolating the ingester from others.

Both *LOTP*'s poetry and footnotes warn that large doses of opium lead to "intoxication, languor, stupor and death" (II. 270), isolating the ingester. Opium's "marble lover" in "iron slumbers" with "glassy eyes" as "fresh horrors seize" the "stiffening limbs, their vital currents freeze" (II. 289–90; 277–78). Overconsumption once or by repeated use results in "emaciated and decrepid appearance" of addicts with "ridiculous and idiotic gestures". *LOTP* therefore advocates careful (scientific) observation and management of opium doses lest Hermes' "dread" caduceus—like the pipe of Pan, the syringe of Syrinx or the apothecary's pestle of any drug—reduce the ingester's productivity and happiness. Darwin's examples of opium addiction, however, live in far-away places rather than close to home, where he perceived alcohol to be the greater health threat. *LOTP* and Darwin's subsequent works were not enough to prevent many in Darwin's orbit, like Coleridge, Wilberforce and De Quincey, from struggling to manage what they called opium "[c]raving" and "enslavement" (Vickers 2015, 337; Fulford 1999, 237). Such pain and suffering, according to Darwin, were not mere personal problems but civic concerns.

Tea and Empathy

Poppy's canto ends with a brilliant cultural critique of British addiction to *camellia sinensis* and the financial profits driving the global and often illicit silver, tea and opium trade (Flynn and Giraldez 2002; Chen 2017;

[2] See essay in this volume by Anna Rowntree.

Merritt 2017). One of the "drug foods" of Europe's "caffeine revolution" (Jamieson 2004, 287; 285), tea, along with coffee and chocolate, was one of the major non-European commodities regularly consumed by the English. Although the East India Company initially "struggled to stimulate consumer demand for the relatively novel commodity", by the 1750s tea habits had expanded throughout the British Empire so that the once-luxury item was an "everyday [...] necessity" across socio-economic classes (Merritt 2017, 5). In 1785, English tea imports doubled those of coffee (Jamieson 2004, 284), and tea-drinkers were increasingly dependent on "links to credit, services, and goods from India, Asia, and the Pacific beyond" (Merritt 2017, 4) to maintain their access to this "building block" of their eighteenth-century imperial economy (Merritt 2017, 3–4).

At the end of *LOTP*'s sedating Canto II, Hygeia, working with Eastern opium, encounters "Philanthropy" in Darwin's homage to public health heroes, such as contemporary prison reformer John Howard, an important symbolic association of imprisonment, slavery, intoxication and mental health. After arriving at Hygeia's shrine of medicinal herbs (Harris 2005, 323–28), Hygeia and Philanthropy abruptly pause their civic activities to observe teatime, so mesmerised are they by paradoxically stimulating yet soothing substance and the ritual social performances around its consumption.

Darwin's tea-drinking subjects stare vacantly into "gaudy cups" of "steamy treasure" enervated rather than innervated, despite the stimulating effects of "the green herb of China's envy'd bowers" (II. 481–82). Rejecting nationalistic notions of English nervous sensitivities as a sign of unique virtue (Vickers 2015, 328–29), Darwin suggests that British tea-drinkers, like opium-eaters, could become hypnotised and civically paralysed by their addiction to the pleasures and profits of the colonial economy. The twelve poetic lines about tea have no accompanying explanatory note, which is unusual in *LOTP*. But in the subsequent Interlude II, the Poet wonders to the Bookseller how to rouse reader-consumers from laudanum-like apathy to greater sympathy for those suffering the underside of the global economy. Darwin insists the reader imagine one's own health in relation to others' wellbeing.

LOTP's structural and philosophical centre occurs here in the middle of three interludes, between Canto II's sedating reveries and Canto III's frightening poisons. Here, Darwin's fictive Bookseller and Poet dialogue about classical and contemporary concepts of pleasure, pain, thrill and relief. Before nineteenth-century philosophers and proto-psychologists named the process of empathy (recognising resemblances and realising connections among beings), *LOTP*'s exploration of Mark Akenside's 1768 "Pleasures of the Imagination" and Lucretius' classical shipwreck pinpoint Darwin's "strongly Lucretian sense of poetic purpose [...] formulated [...] as early as 1781" to emulate Lucretius' introduction of Epicurean ideas through didactic poetry. Anti-Jacobins accused Darwin of writing *LOTP* merely for erotic titillation or a "modernized Epicureanism" with the highest goal of humanity simply the pursuit of pleasure and relief from pain (Jackson 2009, 179–91). Yet, in a twist on Lucretius' take on Epicurus, *LOTP* demonstrates how pain and/or pleasure connects or isolates individuals and creates a civic role for drug therapy as a public good and not merely for personal pleasure projects.

LOTP reflexively cultivates an inner seeing into other creatures' experiences and points of view, through either literary imagination or literal ingesting of their essences. Darwin later allowed that among countless varieties of pleasures and pains there might be "amiable insanities" or "meritorious ones" arising from "happy hallucinations" (Budge 2007, 298). In *LOTP*, sensitivity to the pain of others is an important social stimulant for the wellbeing of society. Our own relief at having escaped personal pain should not become indifference to others' suffering or personal shipwrecks. The point of stimulating readers' reason and imagination is to generate empathy for others' pain and pleasure as well as to understand our own. *LOTP*'s "clear social imperative" is civic awareness and involvement (Jackson 2009, 188–89), evident in its critiques of the colonial economies of opium, tea, alcohol and slaves.

Monsters, Madness, Slavery and Alcohol

British tea consumption increased with the sugar trade reliant on enslaved labour, and merchants "involved in [...] intercolonial commerce probably shipped tea and slaves together" (Merritt 2017, 5). Whereas the opiate reverie-like Canto II concludes with slowly steeping British addiction to fruits and drugs of imperialism, Canto III—Darwin's most pointed commentary on psychoactive drugs—opens with a gothic physic-garden-turned-cemetery where "shapeless spectres quiver", bats "scream", dogs howl and "all tremble" (III. 40–43). It is a ruin of parasites, poisons, lunatics, slaves, enslavers and wine.

Although "[d]runkenness among the wealthy was often described as a private vice, and while drunkenness and addiction to alcohol among the labouring poor were said to pose 'political mischiefs'" (Rabin 2005, 457), Darwin believed that alcoholic beverages posed the greatest potential threat to individual and collective wellbeing across the socio-economic spectrum. This was not because alcohol was inherently more lethal than enchanter's nightshade or cherry laurel found in Canto III, but rather due to alcohol's accessibility and integratedness in informal and formal economies, in addition to human adaptation to alcohol's effects through habitual, increased consumption over time. In other words, for Darwin alcohol carried the biggest risk for what would later be called addiction.

Eighteenth-century Europeans experimentally fermented and distilled almost "everything the earth produces, flowers, fruits, seeds, spices, aromatic and vulnerary plants" (Spary 2012, 151–53). Measuring and assessing the "proof" of alcohol content were difficult due to the properties of alcohol, water and other added elements for flavours; it was often contested among scientists, merchants, guilds and government officials (Ashworth 2001, 32; 45). Citing instances of sudden deaths and speculating about the unregulated processes and experiments with fermentation and distillation, Darwin issued an important warning about the toxicity of raffias, a popular new class of spirits distilled from whole pitted fruits (Spary 2012, 154). *LOTP* compares these liqueurs to cherry laurel extract, "the most sudden poison we are acquainted with" (III. 40), because they produce "intoxication or inspiration" like that of a Pythian priestess delivering oracles in altered states of consciousness. Whatever enlightenment these biotics enabled would not be worth the danger they posed. While cherry laurel was already considered a poison, Darwin asserted that raffias also offered only a "feast malignant" of "poison in the nectar'd bowls" (III. 40; 335–80).

Other alcoholic beverages and tinctures appear throughout *LOTP*. Mead, for example, appears twelve times, playing on Linnaeus' designation of the *dodecatheon* ("twelve gods") for the *meadia* plant, commonly named for physician Richard Mead. *LOTP* also plays with the term for pastoral grasses and fermented honey drinks as an intoxicating/intoxicated "Cherub-train" of thought roams "from mead to mead" throughout the cantos gathering ingredients for a Sweetgale Drink of myrica gale buds, harebell and primrose. Depending on other ingredients added for flavouring and potency, and depending on the imbiber's physiology, meads in *LOTP* can cause a "light step" and the feeling that "wheels on high" like a beetle suspended by "the gilded fly" (I. 86; IV. 60–64; 73–77; 83–92; 145–88) or "Gay hopes" or "amorous sorrows" (1. 5–6). Like the "Rolls" and "waves" (I. 60–64) of *meadia*'s morphology, brewing and psychoactive effects of avenas may be stimulating and/or depressing; they ambiguously mix revelry and reverie.

Fermented meads may have medicinal value, especially when flavoured with mildly narcotic clove, one of Darwin's favourite ingredients, and they may strain an impotent "bow reluctant to its string" and "spread the tender growths of culture'd love", but imbibing too much of them all at once or too often can "letter'd lines decay" and "Love's soft forms neglected melt away". Similarly, distilled "fleecy squadrons" of wheat, barley, rye and oats might have iatric uses, such as in tinctures. But like bubbling, brewing sulphuric hot-spring spas, distilled spirits can be unpredictable: "Impetuous" volcanic eruptions occur when "imprison'd waters boil". Like the brewery, a drunkard can become a "flaming furnace" and "Turn with hot tongs" a "dreadful dart" (IV. 176–84). While plain grape juice might be a "nutritive and agreeable food", wine is a "poison" comparable to the most lethal drugs (III. 355) that dip the writer's "point in poison for the mind" (IV. 157–64). It lessens rather than increasing mental, sexual and social productivity.

Citing his own "[t]reatise on the inverted motions of the lymphatic system" and "[t]heory of the Diabætes and Dropsy, produced by drinking fermented or spirituous liquors", Darwin blamed excessive alcohol drinking for "more than half of our chronical diseases" in the West. Despite the fact that repeated alcohol consumption, like opium with which it was often paired, made diagnosis of underlying conditions more difficult as secondary disorders developed, *LOTP* links "leprous eruptions on the face, gout, dropsy, epilepsy, insanity" to alcoholism (III. 355). The didactic poetry implies a parallel to the 1665 London Plague's "gasping" crowds leaving a wake of wailing, bereft mothers. It dramatises Darwin's concern that "all the diseases from drinking spirituous or fermented liquors are liable to become hereditary, even to the third generation; gradually increasing [...] till the family becomes extinct" (III. 369–87). With the future of humanity at stake, Darwin, although generally an advocate of self-determination, recommends in *LOTP* greater regulation of alcohol production and consumption (III. 387).

Darwin's much-discussed dosing of his first wife and her self-medicating with alcohol and opium both alleviated and intensified her ailments before she died at age thirty (Fara 2012, 23). Mary Darwin's frequent paroxysms in "delirium" and "hysteric fits" undoubtedly inspired some of *LOTP*'s vivid nightmare passages and explanations of what her husband believed "allied to Epilepsy" (Cook 1996, 59). Although Mary's suffering was "sometimes relieved by great doses of opium, and some wine", her frequent alcohol consumption in such "great quantity" to the point of "intoxication" led to cirrhosis of her liver (Cook 1996, 59). Evoking images of Fuseli's *Nightmare* that opens Canto III, one who "swallows drams" and takes the "spirituous" fire inside the pulmonary and hepatic systems becomes—or succumbs to—the incubus, a "gluttonous bird, exulting in his pangs" and tearing "his swoln liver with remorseless fangs" (III. 369–78). Although he rarely explains his classical references in *LOTP*, Darwin elaborates on his Promethean image for alcoholic beverages, which he repeats in later works, "so apt an allegory for the effects of drinking spirituous liquors" (III. 369) is this "disease" of "Drunkenness" (*Zoonomia* XXI). Like maniacs and brutes without volition, Prometheus' "writhing limbs in vain he twists and strains / To break or loose the adamantine chains" (*LOTP* III. 335–80).

Those considered "mad" in eighteenth-century Western societies were alternately feared, pitied and even envied for their unrestrained abnormal behaviours (Scull 1993, 64). Uncontrollable behaviour in "mad" George III's society was commonly met by chains and corporeal abuse in homes, asylums and prisons (Cherry 2003, 24–25), which Darwin worked to change. Writing in a newly "broad 'neuropathological' paradigm" about "proto-neurology" (Vickers 2015, 327–29), Darwin, arguably the "most important European psychologist" at the end of the century (Reed 1997, xi), sought to educate readers out of ignorance, apathy and cruelty regarding intoxication and little understood neurological disorders, which he viewed as related. *LOTP* thus introduced readers to Darwin's theories of neurophysiology and mental health (Budge 2007, 279), including potential drug therapies and paraphernalia (pipes, chimneys, syringes, etc.) as well as Darwin's drug-free therapeutic experiments with electricity and mechanical devices intended to simulate psychostimulants' sublime principle in which initially painful sensations of shock, dizziness or disorientation eventually become pleasurable through repetition and de-sensitisation (see Elliott 2008; Breathnach 2010).

Darwin viewed intoxication and little understood neurological disorders as related because they involved the mind's "perception" of the "nerves of sense" during sleep, sleep-walking, delirium or inebriation (III. 74; 18; 3. 67; 3. 74). Darwin's essay "The Will presides not" in Canto

III introduces his *sensorium* theory found in his later works and explains *LOTP*'s allusions to threads and "kindred wire" like musical chords (I. 97–107) between the mind and nerves and between animals and plants. Analogising "wild" or "mad" (III. 131–78) animal/human physiology to the "automaton"-like morphology and mechanistic movements of impatiens (a plant traditionally associated with insanity), *LOTP* portrays involuntary or compulsive movements as prompted by a variety of possible causes: neurological disorders, inebriation or such psychological trauma as the death of a child, abuse or slavery (III. 131–78).

Illustrating sympathy with and compassion for individuals struggling with volition and involuntary movements, *LOTP*'s Ovidian "short episode" (Inter II) and "rigorously end-stopped Popeian heroic couplet" (Priestman 2013, 37) evoke feelings of innervated mental activity, such as intoxication. Stream-of-consciousness episodes run indistinguishably from one another, and "near-manic" couplets "bounce with exuberance" that can be "thoroughly disconcerting to the reader" (Mahood 2008, 55–56). Intoxication, then, is a state like sleep, epilepsy or insanity: a "deficiency in the power of volition" (Page-Jones 2018, 88). Earlier in the century, "drunkenness" had become increasingly recognised as a mitigating factor in judicial considerations of severe criminal punishment, which Darwin was interested in reforming. Alcohol's effects were now understood to interfere with "moral judgment and self-restraint" (Rabin 2005, 458), and "[r]ather than absolute guilt, the plea of drunkenness might place one closer to the excusable end without implying innocence" (Rabin 2005, 477). For Darwin, drunkenness was a prison of its own, less a comment on the morality of the individual imbiber than an indictment of the civic economy.

Darwin criticised individual drinkers less than those merchants and government officials who profited from others' enslavement to alcohol "under the names of rum, brandy, gin, whiskey, usquebaugh, wine, cyder, beer, and porter" economies (*Zoonomia* XXX. III). Echoing Jeanne-Marie LePrince De Beaumont's 1756 adaptation of *Beauty and the Beast*, *LOTP* contrasts public fascination with the grotesque and lurid "amusement" of popular horror stories with what Darwin called "the bitter cup of true Tragedy". *LOTP* identifies "real misery" as "drunkards and diseases", dying children, grieving mothers, misunderstood neurological disorders and slavery. The real "monsters" are those who inflict or fail to stop the suffering of others, whether in the practice of slavery or the alcohol economy.

LOTP associates consumers' new and old cravings for alcoholic spirits with those drunk on the financial fruits of imperial enslaved labour. A "maudlin" effect of the "seductive" wine (*LOTP* III. 355–56) is the transformation of humans into brutes, physical and/or political incubus, who like parasitical vines enjoy their pleasures at others' expense, strangling them into graves as their mothers go insane from the grief (III. 415–34). Personifying cinnamon, cocoa and other commodities from Britain's American and Asian colonies (Merritt 2017, 4), Darwin depicts the vulnerability, exploitation and grief of children ("emigrant seeds") in the "curst Slavery" of "Oppression's hands" perpetrated by Europeans addicted to pleasure at others' expense (III. 456; 435–36). Canto III culminates in an encomium to abolitionist (and opium-addict) Wilberforce and attacks those who participate in slavery's collective madness on which their colonial economy relied (III. 415–58). Referring to alcohol as the "curse of the Christian world" (III. 355) like slavery, Darwin implies by parallel that European capitalists who incentivise, profit from or simply ignore the perils of overconsumption of alcohol are perpetuating multiple forms of slavery, and anyone "WHO ALLOWS OPPRESSION, SHARES THE CRIME" (III. 456).

Conclusion: A Psychoactive Civics Experiment

In *LOTP*, if drugs could cause undesired behaviours, they could also alleviate them. As pleasurable and painful apperceptions stalk each other "on the flowery brink" of human self-consciousness (IV. 299–320), "toxics" are either a cause of or remedy for undesired experiences. The final canto resolves strands of danger/fear/poison with thrill/relief/healing in an ecstatic, synesthetic sublime, signalling Darwin's cautious optimism about pharmacological progress towards improved individual health and collective eudemonia. As species inhabit and metabolise each other, they transmogrify until it is difficult to distinguish a Pard from a Swan, a nightingale from a rose, a marigold from a cuckoo, a lord from a lady, a beast from a beauty. New "mules" and other "combining forms" emerge from human-plant interactions as *LOTP* imagines a sort of posthumanist species (IV. 299–320). The image of compounding, stimulating Proteus-Love and ironically soothing Devil's Claw offers a final, perfect psychopharmacological pun as a *swallow* (IV. 299–320).

Mocked by some contemporaries as dizzying like an alkaloid itself (Fara 2012), *LOTP*'s reflexive, self-conscious elements give the work an "almost postmodern feel" now (Priestman 2013, 55). Dahlia Porter notes, however, that *LOTP*'s pedagogical associations and analogies are not unbounded. Darwin's use of heroic couplets and end stops, typographical divisions between poetry and prose on the same page and his multiple taxonomic orders counterbalance a sense of drugs' indeterminacy. Rather than portraying the physic garden as hopelessly unknowable, *LOTP*'s "relationship between empirical science and imaginative art" optimistically balances the tension "between the two organizing topi, taxonomy (the drive to distinguish, separate, and compartmentalize)" and "analogy (the desire to connect and relate)" (Porter 2007, 213–14).

Darwin's conservative nemesis, the *Anti-Jacobin*, warned writers influenced by *LOTP* "to watch the inroads of science [...] lest the intermixture of scientific discovery with poetic invention should become fashionable". *The Critical Review*, however, praised it for realizing Darwin's goal that "what at first began in amusement, may terminate in scientific acquisition" (Jackson 2009, 193). *LOTP* increased humans' imagination and understanding of drugs' capacity to effect or stymie an "evanescent" energy rhapsodised as the "Sweet Echo" of the Æolian lyre (*LOTP* IV. 348; 350; I. 97–107) resonant through Coleridge's *Eolian Harp*, Keats' *Endymion* and Shelley's *Ode to the West Wind*. *LOTP* expressed sympathy for various reasons humans choose or are compelled to ingest drugs for relief, stimulation or enhancement; and it urged accountability for politico-economic decisions and systems that exploit or enslave others. *LOTP*'s conjugation of pleasure and profit, its blend of erotic imagination and rationalism helped to fan the "fire from which 'Romanticism' sprang to life [...] not so much at the urging of Wordsworth and Coleridge but from [...] Darwin's legacy" (Jackson 2009, 179).

Darwin died in 1802 as admirers and critics, all desperate for more explicit taxonomies of helpful and harmful drugs, began seeking more regulation of drugs' production and consumption. Despite his complicated relationship with Darwin, Coleridge in "Essays on Method" (1818) praised him for improving ideas about the "relation of things" (Porter 2007, 213–14; Ullrich 1984). Beddoes continued Darwin's public health advocacy by educating general readers in pharmacognosy, observing in *Hygëia: or Essays Moral and Medical* (1802–03) that "there are no means of promoting morality upon which we can rely, except the knowledge of the true relations between man and other beings or bodies" (Grinnell 2006, 225–27). Such pursuits ostensibly aimed at the kind of individual and collective wellbeing envisioned by Darwin, but Romantic and

Victorian notions of an "economy of the self" (Budge 2007, 284) and "moral economy of health" (Grinnell 2006, 236) increasingly objectified, commodified and isolated plants and humans alike.

Reading *LOTP* at the end of the eighteenth century was more than a game to benefit a singular reader; interaction with "toxics" should enhance meaningful connections of as many beings as possible rather than isolating them. Contrary to his critics' claims, Darwin's idea of health and happiness was more than momentary stimulation or relief from acute pain. Braiding the forces of Hermes (medicine) and Aphrodite (love/compassion), *LOTP* imagined individual drug gambolers incorporated into an economy of civic care and empathy rather than a parasitical society of self-centred or apathetic drug consumers and exploiters. Anticipating Darwin's last poem, in which humans "should eye with tenderness all living forms" (*TON* IV. 427–28), *Loves of the Plants* contributed to a "science of drugs that originate from living beings and are studied to help other living beings" (De Pasquale 1984, 16) in addition to one's self.

References

Ashworth, William J. 2001. Beneath the Trader and the Public': British Alcohol Standards and the Proof of Good Governance. *Technology and Culture* 42 (1): 27–50.

Barfoot, Michael. 1988. Brunonianism under the Bed: An Alternative to University Medicine in Edinburgh in the 1780s. In *Brunonianism in Britain and Europe*, ed. W.F. Bynum and Roy Porter, 22–45. London: Wellcome Institute for the History of Medicine.

Barker-Benfield, Graham John. 1992. *The Culture of Sensibility: Sex and Society in Eighteenth-Century Britain*. Chicago: University of Chicago Press.

Boon, Marcus. 2002. *The Road to Excess: A History of Writers on Drugs*. Cambridge: Harvard University Press.

Breathnach, Caoimhghín S. 2010. Hallaran's Circulating Swing. *History of Psychiatry* 21 (1): 79–84.

Browne, Janet. 1989. Botany for Gentlemen: Erasmus Darwin and 'The Loves of the Plants'. *Isis* 80 (4): 593–621.

Budge, Gavin. 2007. Erasmus Darwin and the Poetics of William Wordsworth: 'Excitement without the Application of Gross and Violent Stimulants'. *British Journal for Eighteenth-Century Studies* 30: 279–308.

Chen, Song-Chuan. 2017. *Merchants of War and Peace: British Knowledge of China in the Making of the Opium War*. Hong Kong: Hong Kong University Press.

Cherry, Steven. 2003. *Mental Health Care in Modern England: The Norfolk Lunatic Asylum, St Andrew's Hospital, 1810–1998*. Martlesham: The Boydell Press.

Cook, Gordon C. 1996. Mary Darwin's Illness. *Notes and Records of the Royal Society of London* 50 (1): 59–63.

Courtwright, David T. 2001. *Forces of Habit: Drugs and the Making of the Modern World*. Cambridge: Harvard University Press.

Darwin, Erasmus. 1794–1796. *Zoonomia; or, The Laws of Organic Life*, 2 vols. London: J. Johnson.

———. 2007. *The Botanic Garden: Part II—Containing The Loves of Plants. A Poem. With Philosophical Notes*. Echo Library.

DerMardersian, Ara H. 1995. Milestones of Pharmaceutical Botany. In *Symposium Proceedings, American Institute of the History of Pharmacy Annual Meeting*.

De Pasquale, Anna. 1984. Pharmacognosy: The Oldest Modern Science. *Journal of Ethnopharmacology* 11 (1): 1–16.

Dugan, Holly. 2011. *The Ephemeral History of Perfume: Scent and Sense in Early Modern England*. Baltimore: Johns Hopkins Press.

Elliott, Paul. 2008. 'More Subtle than the Electric Aura': Georgian Medical Electricity, the Spirit of Animation and the Development of Erasmus Darwin's Psychophysiology. *Medical History* 52 (2): 195–220.

Fara, Patricia. 2012. *Erasmus Darwin: Sex, Science, and Serendipity*. Oxford: Oxford University Press.

Flynn, Dennis Owen, and Arturo Giraldez. 2002. Cycles of Silver: Global Economic Unity Through the Mid-Eighteenth Century. *Journal of World History* 13 (2): 391–427.

Foucault, Michel. 1994. *The Order of Things: An Archaeology of the Human Sciences*. New York: Vintage.

Fulford, Tim. 1999. *Romanticism and Masculinity: Gender, Politics and Poetics in the Writings of Burke, Coleridge, Cobbett, Wordsworth, De Quincey and Hazlitt*. New York: Macmillan.

Green, Matthew. 2007. Blake, Darwin, and the Promiscuity of Knowing: Rethinking Blake's Relationship to the Midlands Enlightenment. *Journal for Eighteenth-Century Studies* 30 (2): 193–208.

Grinnell, George C. 2006. Thomas Beddoes and the Physiology of Romantic Medicine. *Studies in Romanticism* 45 (2): 223–50.

Harris, Stuart. 2005. The Poet as Pathologist: Myth and Medicine in Erasmus Darwin's Doctrines of Love and Imagination. In *The Genius of Erasmus Darwin*, eds. C.U.M. Smith and Robert Arnott, 321–36. Surrey: Ashgate.

Jackson, Noel. 2009. Rhyme and Reason: Erasmus Darwin's Romanticism. *Modern Language Quarterly* 70 (2): 171–95.

Jamieson, Ross W. 2004. The Essence of Commodification: Caffeine Dependencies in the Early Modern World. *Journal of Social History* 35 (2): 269–94.

Johnson, Victoria. 2018. *American Eden: David Hosack, Botany, and Medicine in the Garden of the Early Republic*. New York: W. W. Norton.

Komisaruk, Adam. 2014. Mystifying What Matters: Erotic Antiquarianism in Erasmus Darwin's Portland Vase. *Eighteenth-Century Life* 38 (3): 1–29.

Kuhn, Thomas. 1971. The Relations Between History and History of Science. *Daedalus* 100 (2): 271–304.

Lawrence, Christopher. 1988. Cullen, Brown and the Poverty of Essentialism. In *Brunonianism in Britain and Europe*, ed. W.F. Bynum and Roy Porter, 1–21. London: Wellcome Institute for the History of Medicine.

Maehle, Andreas-Holger. 1999. *Drugs on Trial: Experimental Pharmacology and Therapeutic Innovation in the Eighteenth Century*. Amsterdam: Editions Rodopi B.V.

Mahood, Molly Maureen. 2008. *The Poet as Botanist*. Cambridge: Cambridge University Press.

McKertich, Padma V., and V. Shilpa. 2016. 'It Happens Quietly': Plant Poetry and the *Botanification* of the Imagination. *Journal of Literature and Science* 9 (2): 36–49.

Merritt, Jane T. 2017. *The Trouble with Tea: The Politics of Consumption in the Eighteenth-Century Global Economy*. Baltimore: Johns Hopkins University Press.

Page-Jones, Kimberly. 2018. The 'Strange Self-Power in the Imagination': Epilepsy, Fancy and Disbelief in the Writings of Coleridge, Erasmus Darwin and Thomas Beddoes. *The AnaChronisT* 18 (1): 83–100.

Paine, Thomas. 1870. To the Editor of the Pennsylvania Magazine, 1775, 'The Snowdrop and Critic, A Dialogue'. In *The ThÆlogical Works of Thomas Paine*. J.P. Mendum Investigator Office.

Porter, Dahlia. 2007. Scientific Analogy and Literary Taxonomy in Darwin's *Loves of the Plants*. *European Romantic Review* 18: 213–21.

Porter, Roy. 1982. Mixed Feelings: The Enlightenment and Sexuality in Eighteenth-Century Britain. In *Sexuality in Eighteenth-Century Britain*, ed. Paul-Gabriel Boucé, 1–27. Manchester: Manchester University Press.

Priestman, Martin. 2013. *The Poetry of Erasmus Darwin: Enlightened Spaces, Romantic Times*. Surrey: Ashgate.

Rabin, Dana. 2005. Drunkenness and Responsibility for Crime in the Eighteenth Century. *Journal of British Studies* 44 (3): 457–77.

Reed, Edward S. 1997. *From Soul to Mind: The Emergence of Psychology from Erasmus Darwin to William James*. New Haven: Yale University Press.

Robinson, Eric. 1954. Erasmus Darwin's Botanic Garden and Contemporary Opinion. *Annals of Science* 10: 314–20.

Scull, Andrew. 1993. *The Most Solitary Afflictions: Madness and Society in Britain, 1700–1900*. New Haven: Yale University Press.
Spary, E.C. 2012. *Eating the Enlightenment: Food and the Sciences in Paris, 1670–1760*. Chicago: University of Chicago Press.
Taylor, Mark. 2011. Planting for Pleasure: The Eighteenth-Century Erotic Garden. *Interiors: Design, Architecture, Culture* 2 (3): 357–70.
Ullrich, David W. 1984. Distinctions in Poetic and Intellectual Influence: Coleridge's Use of Erasmus Darwin. *The Wordsworth Circle* 15 (2): 74–81.
Valsania, Maurizio. 2004. 'Another and the Same': Nature and Human Beings in Erasmus Darwin's Doctrines of Love and Imagination. In *The Genius of Erasmus Darwin*, ed. C.U.M. Smith and Robert Arnott, 337–56. Surrey: Ashgate.
Vickers, Neil. 2015. Opium as a Literary Stimulant: The Case of Samuel Taylor Coleridge. *The International Review of Neurobiology* 120: 327–38.

CHAPTER 6

Pharmacokinetics and Opium-Eating: Metabolites, Stomach Aches and the Afterlife of De Quincey's Addiction

Hannah Markley

In 2008, opioid overdoses superseded automobile accidents as the leading cause of accidental death in the United States.[1] This turning point defined the terms of the contemporary opiate epidemic, precipitated by the proliferation of heavily marketed prescription painkillers, like OxyContin and Fentanyl, in the 1990s. Ironically, nearly two-hundred years after *Confessions of an English Opium-Eater* (1821) first appeared, these pharmaceuticals gave new material forms to the term 'opium-eating': Fentanyl is available as a lollipop (Actiq), and the warning label on the OxyContin prescription bottle instructs that the pill "should never be

[1] This statistic is drawn from Sam Quinones book *Dreamland: The True Tale of America's Opiate Epidemic* (x), published in 2015.

H. Markley (✉)
Georgia Institute of Technology, Atlanta, GA, USA
e-mail: hannah.markley@lmc.gatech.edu

N. Roxburgh and J. S. Henke (eds.), *Psychopharmacology in British Literature and Culture, 1780–1900*, Palgrave Studies in Literature, Science and Medicine,
https://doi.org/10.1007/978-3-030-53598-8_6

chewed" because chewing the pill nullifies its continuous release mechanism, releasing potentially deadly amounts of oxycodone.[2] Opium-eating, of course, has a long history in British literature, and the writings of Thomas De Quincey, the most infamous opium-eater of the nineteenth century, can help reframe the terms of opiate use and abuse.[3] In De Quincey's text, the metaphor—or, as we shall see, partial-metaphor—of 'opium-eating' marks the intersections of different discourses (medical, literary, moral and physiological) as they are deployed to describe opium's drug action and effects in different registers, from confession to scientific deduction. By mobilising 'opium-eating' as a figure for these intersections, this essay argues that opiate use can be reconceived through the co-constitution of discursive modes, emphasising the ways that opioid dependence results from complex interactions among the mind, body and drug that cannot be understood as the result of a drug acting unilaterally on the body.

This essay explores De Quincey's 'opium-eating' alongside psychopharmacological explanations of how the body absorbs, distributes, metabolises and excretes opium. In the early nineteenth century, medical theory derived from the work of John Brown, viewed the body as a conduit of excitability that could be stimulated by internal (thinking, emotion, passion) or external (food, wine, opium) forces. Over—or under—stimulation resulted in illness. In Brown's system, external stimuli, like opium, were used to generate excitation in the medullary nerves and muscles where human vitality was presumed to reside.[4] In this respect, De Quincey describes opium as a 'stimulant' because he

[2] Quinones, noting the irony of the pharmaceutical drug, suggests that addicts learned how to abuse the drug from its warning label, often chewing, crushing and snorting the drug. Additionally, the drug is also frequently dissolved in water and injected (Quinones 2015, 120–26).

[3] In his book *The Pursuit of Oblivion* (2004), Richard Davenport-Hines describes how the practice of literal opium-eating in the nineteenth century was primarily associated with colonial subjects in India and Turkish customs, where opium was extracted from the opium poppy *Papver soniferum*, moulded into lumps, cakes or bricks, and processed by boiling it to remove impurities. The processed opium is the consistency of a dense modeling clay, which was either incorporated into spiced cakes or smoked (Davenport-Hines 2004, 29–30).

[4] In "Coleridge, Thomas Beddoes, and Brunonian Medicine" Neil Vickers details the development of the Brunonian system, its influence on Beddoes' *Hygeia,* and its relation to Coleridge's commentaries on German idealism (1997, 65–68). See also George Grinnel's essay "Thomas Beddoes and the Physiology of Romantic Medicine" (2006, 223–50).

understands its drug action as exciting the nervous tissues of his body. My approach shifts focus from paradigms that focus primarily on what opiates do to the body to emphasise what the body does to the drug.[5] In modern psychopharmacology, drug action is defined as a drug's mechanisms of action on the brain and body, while drug effects are the downstream physiological expressions of these actions. Psychopharmacology considers these two processes as the pharmacodynamics of a drug. Drawing on Elizabeth Wilson's arguments about the significance of digestive organs and the peripheral body for the production of psychoactive effects in the central nervous system, known as pharmacokinetics, I argue that the term opium-eating forces us to reckon with the role of digestive and metabolic processes in accounts of opiate use. Moreover, even as De Quincey's account of the pains of opium, as Robert Morrison observes, is contradicted by modern pharmacology, renewed attention to 'opium-eating' as a figure reveals the ways that De Quincey's withdrawal, like the Brunonian system through which he seems to understand it, entangles his habit with its symptom, the drug with its additives and his central nervous system with organs, actions and feelings in his peripheral body, including his stomach, intestines and liver.[6]

For De Quincey, the term opium-eater was always a semi-figurative nom de plume; he, like other opium users in the early nineteenth century, mostly drank laudanum—a tincture of opium dissolved in alcohol. At the height of his opium use, De Quincey claimed he consumed "twelve thousand drops of laudanum a day", a large dose for a man under five feet tall (Morrison 2009, 271).[7] But 'opium-eating' was not only a figure for the consumption of this solution. J. G. Lockhart, a novelist, essayist

[5] In a chapter entitled "Chemical Transference", from her book *Gut Feminism* published in 2015, Elizabeth Wilson argues that in order to understand the clinical effects of SSRI medications on depression, we must pay closer attention to pharmacokinetic processes (digestion, distribution, metabolisation and elimination of a drug) as opposed to the pharmacodynamic processes (how the drug affects the body) (Wilson 2015, 100–05).

[6] For an account of the relationship among cognitive science, the activities of the central nervous system, and De Quincey's representations of the 'unconscious', see Markus Iseli's book *Thomas De Quincey and the Cognitive Unconscious,* published in 2015.

[7] In his 2011 article, drawing on Barry Milligan's book Pleasures and Pains: Opium and the Orient in Nineteenth-Century British Culture (1995), Morrison observes that it is difficult to translate these drops into an actual dose of opium because concentrations of opium in preparations of laudanum varied as well as the potency of the opium procured for these tinctures (Morrison 2009, 271; Milligan 1995, 22–27).

and biographer of Sir Walter Scott, reported in 1815 that De Quincey "drank all [the alcohol] he could get and in between glasses [of wine] kept munching opium pills out of a snuff box as another man might munch filberts" (Lockhart quoted in Morrison 2009, 271). Yet while De Quincey consumed shocking amounts of opium in laudanum and pill form, he seldom describes actual eating in his texts (Burt 2009, 157). He suffered from stomach pains all his life and was a fitful eater at the best of times. As Paul Youngquist reminds us, "De Quincey's largely liquid diet was an attempt to appease his bowels", fasting, De Quincey claimed towards the end of his life, "for thirty years" (Youngquist 1999, 356). In this sense, the image of De Quincey chewing on opium pills as if they were hazelnuts materialises the term opium-eating in ways that, like crushed up OxyContin pills or Fentanyl lollipops, poses important questions about the relationship between opiate consumption and digestive processes. If De Quincey's term opium-eating reflects the cyclical exigencies of opiate use and abuse, it is worth reading the term through the physiological imbrication of opium use and stomach pain. As Youngquist shows, De Quincey ate opium to alleviate "symptoms caused by removing it from his diet", a paradox that reflects the tenants of the Brunonian system of medicine in which a substance can simultaneously operate as a cause of ill health and its remedy (Youngquist 1999, 356). In this way, the entanglement of opium and digestive health becomes a signifier of the displaced relationship between cause and effect in modern paradigms of addiction. While addiction as a concept did not yet exist in the early nineteenth century, in Thomas Beddoes' introduction to his translation of Brown's *The Elements of Medicine*, he argues that Brown died because "he swallowed as he went to bed a very large dose of laudanum; a species of dram to which he, indeed, had long been *addicted*" (Brown 1795, xcii, emphasis added). In this respect, Beddoes yokes the term addiction not only to opium but also to the Brunonian system in which the displaced relationship between cause and effect produces opium as both the cause of ill health and its only possible cure.[8]

The interrelation of De Quincey's opium habit with digestion as figurative, physiological and pharmacological processes, in turn, reveals how the physiological and chemical dependencies o f φpiate use participate in the discursive production of medical, moral, emotional and literary effects.

[8] See essay in this volume by Anna Rowntree.

This essay explores the discursive and physiological entanglements at work in early nineteenth-century medicine alongside modern psychopharmacology to argue that De Quincey locates his stomach as the epicentre of his habit by insisting on the imbrication of brain and gut, drug action, effect and side effect. The essay concludes by asking how 'opium-eating' taken in this sense might reframe our understanding of the contemporary opioid epidemic. While "The Opium-Eater", as he was known in *Blackwood's Magazine*, does not offer solutions to opioid addiction, the image of an almost child-like man chewing on opium pills in between glasses of wine casts a shadow into the twenty-first century.

Opium-Eating and the Pharmacokinetics of Laudanum

In order to understand the significance of 'opium-eating' for psychopharmacology, we need to recognise the significance of the stomach as trope in De Quincey's writing, nineteenth-century medicine and in the literary criticism of his work. Critics such as Youngquist and David L. Clark have explored De Quincey's accounts of his body and the ways he situates the stomach in relation to Beddoes' *Hygeia* and philosophies of mind, respectively.[9] For these critics, his 'opium-eating' reflects and produces digestive tropes that De Quincey circulates throughout his work as extended metaphors for the co-implication of psychoactive drug use, Romantic imagination and "a habitual well-being that is contingent and variable", rooted in bodily maintenance (Youngquist 1999, 356). Clark helpfully situates De Quincey's obsessive readings of Kant in relation to the stomach and its digestive processes, citing a passage from De Quincey's 1834 portrait of Coleridge: "[A]s the stomach has been known, by means of its own potent acid secretions, to attack not only whatsoever alien body is introduced within it, but also sometimes [...] to attack itself and its own organic structure [...] so did Kant carry forward his destroying functions" (Lindop 2000, 156). Following Daniel O'Quinn's essay on digestive figures in De Quincey's writings on Kant,

[9] For more on De Quincey's stomach as a literary and philosophical figure, see Paul Youngquist's "De Quincey's Crazy Body", published in 1999. For a psychoanalytic perspective on De Quincey's psychosomatic body, see Joel Faflak's chapter "De Quincey Terminable and Interminable", from his book, *Romantic Psychoanalysis: The Burden of the Mystery,* published in 2008.

Clark examines the ways in which images of a consumption that consumes itself—of digestive organs that eat themselves—structure De Quincey's understanding of his own "crazy body" as it is figured by the dual actions of opium-eating and metaphysics.[10] While these critics have explored De Quincey's writings on the stomach as both organ and image influenced by Kant, I want to explore how the Kantian topos of a stomach that feeds on itself might be understood through the physiological absorption, metabolisation, distribution and excretion of opium. That is, how might De Quincey's stomach and 'opium-eating' be understood differently if we take into account not only the historical medical theories of Brown and Beddoes, but the psychopharmacological processes governing laudanum consumption? In this way, modern psychopharmacological knowledge can be put into productive tension with historical and literary accounts of 'opium-eating'.

To explain the psychopharmacological significance of 'opium-eating', I want to distinguish, following Elizabeth Wilson, between the pharmacodynamics and pharmacokinetic processes (Wilson 2015, 100). While contemporary accounts of addiction focus primarily on drug actions that work on the brain and body, producing physical dependence (pharmacodynamics), I want to focus on how opiates are digested, absorbed, metabolised and excreted by the body. That is, how the body acts on opiates. As Susan Zieger points out, addict in Latin meant a slave or a bondsman and, as it has become synonymous with substance abusers, denotes a person who is a slave to a drug, and bound to it against his or her own will (Zieger 2008, 6). Addiction as a concept describes, narrates and, at times, dictates expectations about what a drug does to a body through this trope of enslavement, producing a concomitant pharmacological emphasis on drug action as an imperious force that usurps the body's proper functions. That is, when Beddoes claims that Brown was addicted to opium, he implies that Brown's death results from a loss of self-control in which the opium takes over. By focusing on how the body digests, absorbs, metabolises and excretes opium, the term 'opium-eating'

[10] Clark's essay was published in 2003. In "De Quincey's Crazy Body", Paul Youngquist also explores the ways in which descriptions of De Quincey's stomach and his brain overlap and co-constitute one another. For more on De Quincey and Kant, see Daniel O'Quinn's essay "The Gog and Magog of Hunnish Desolation: De Quincey, Kant and the Practice of Death", published in 1997.

can describe pharmacokinetic processes that offer alternatives to addiction as a construct and its tropes of chemical enslavement, that echo and reinforce the racial violence of chattel slavery. If the term addict implies a relation in which an abuser is enslaved by the drug, then the implied pharmacological relation between the user and the drug is unilateral: the drug acts on the body (pharmacodynamics). The slave metaphor inherent in the term addict is concretised then by simplifying or even ignoring the pharmacokinetic processes that also play a role in an opiate's psychoactive actions and their downstream effects.[11] This is not to suggest that opiates are not habit-forming, but rather to call attention to the ways that physiological dependence results from complex interactions between the mind, the body and the drug that are neither predictable from one individual to the next nor the result of a drug acting unilaterally on the body.

If we want to track laudanum's drug action on the body and De Quincey's digestive tropes as well as the role the body plays in producing these physiological and literary effects, we need to begin with its chemical composition: opium and alcohol. A dose of opium must be further subdivided into its six active alkaloids: morphine (10%), codeine (0.5%), thebaine (0.2%), papverine (1%) and noscapine (6%). While morphine, codeine and thebaine are known for the analgesic (painkilling) properties, papverine and noscapine are muscle relaxants (Brunton et al. 2011, 498–99). As Barry Milligan acknowledges, it is very difficult to know the precise amount of opium in De Quincey's laudanum preparations because they often varied in potency (concentrations of opium and its active alkaloids) from one batch to the next (Milligan 1995, 22–27). Yet in any preparation, morphine would have been the most concentrated component of the solution. The analgesic effects of morphine work by repressing the nervous system's physical responses to pain as well as the

[11] Wilson includes a fascinating analysis of the placebo effect in clinical trials of SSRIs, drawing on the use of placebos for pain patients. Wilson describes trials in which patients were told they were put on opiates and in fact received placebos; the analgesic effects were the same. By this account, the effects of opiates are tied up in the psychosomatic powers of placebo which work on the physical body through psychological routes (Wilson 2015, 125–27). So while, as Morrison insists, De Quincey's accounts of withdrawal are at odds with its pharmacological effects (opiate withdrawal is often compared to a bad flu), William Burroughs celebrates De Quincey's account as the first and best book about being a "junky" (Morrison 2009, 273–74). De Quincey's withdrawal becomes a kind of placebo—in Wilson's sense of the term—that influences and produces cultural ideas about opiate consumption and withdrawal to this day.

affective reactions that accompany it, so the drug not only treats physical sensations, but works on the moods, anxieties and emotional fallout of pain. Opiates, like morphine, attach to μ-opioid receptors in the brain, inhibiting the transmission of painful sensations in the spinal cord and central nervous system (Brunton et al. 2011, 495). However, opioid receptors are also located in the enteric system, where they regulate the gastrointestinal tract, inhibiting the muscular contractions in the intestines that facilitate bodily elimination (496). These receptors account for the gastrointestinal side effects of opiate abuse and withdrawal. As substances like morphine bind to the μ-opioid receptors in the gut, they inhibit bowel movement, linking the pain killing properties to cramping in the intestinal tract.

Yet, before opiates have any effect on pain in the central nervous system or intestinal motility, they must be absorbed into the body. Laudanum was swallowed, and like OxyContin or a Fentanyl lollipop, absorbed through the digestive tract. The tincture is digested, moving from the stomach and into the intestines where the alkaloids morphine, codeine and thebaine pass through the gut mucosa. From there, the portal vein carries these alkaloids to the liver where they are broken down. These initial movements from the intestines to the liver are referred to as a "first-past clearance" of the drug (Brunton et al. 2011, 499; Wilson 2015, 102). Crucially, the liver breaks up morphine into two different metabolites: morphine-6-glucuronide and morphine-3-glucuronide. It is morphine-6-glucuronide—the metabolite produced by the breakdown of morphine in the liver—that accounts for drug's painkilling properties. That is, the metabolite which is made by the body's own 'digestive' processes is as potent as morphine itself. As Wilson suggests in her analysis of SSRI metabolites, "the difference between drug effect and side effect, between preliminary metabolism and psychoactivity, between distribution and elimination, are constantly being made and remade" by these metabolic processes (Wilson 2015, 102). Furthermore, the implications of these entanglements are far reaching. No longer is opium an "autocratic agent" that acts on the mind and body imperiously; rather, it is a substance that derives its painkilling properties from the metabolic transformations that the body carries out on it (Wilson 2015, 102). Moreover, the processes by which the body 'eats' an opiate are the processes that determine its potency. Opiates such as OxyContin and Fentanyl have been designed to have different rates of absorption, distribution, metabolisation and excretion that radically alter how much of a drug enters the bloodstream after its first-pass in the liver and how easily the drug, once in the bloodstream, crosses the blood/brain barrier. The blood/brain barrier is a tightly packed group of capillaries that surrounds the brain protecting it from foreign bodies. Drugs like Fentanyl move more easily than morphine across this barrier because Fentanyl is fat soluble and can travel through the capillaries more quickly. As a result, Fentanyl is 130 times more potent than morphine, and, potentially, far more deadly.

In this respect, 'opium-eating' might be understood as a process that varies depending on the ways the body interacts and transforms the opiate it ingests. 'Opium-eating', as De Quincey practiced it, entails complex metabolic transformations that circulate painkilling metabolites and nervous system depressants (alcohol) throughout the body by way of the metabolic organs—intestines, liver, gallbladder and kidneys—that in turn produce opium's drug actions and their downstream effects. In this sense, 'opium-eating' can be understood as a figure for the significance of the body's metabolic processes in the production and entanglement of "effect and side effect, preliminary metabolism and psychoactivity, distribution and elimination" in the enteric nervous system as the intestines quiver and seize in response to the opiate metabolites that circulate through the body (Wilson, 102). These pharmacological imbrications of mind and body, effect and side effect, can in turn help us interpret the digestive figures in De Quincey's writings and their ramifications for the ways in which he frames opiate use and withdrawal in literary discourse.[12]

Metabolites in Confessions of an English Opium-Eater (1821)

Rather than claiming that psychopharmacology determines how we understand De Quincey's addiction, I want to use the psychopharmacological implications of the term opium-eating to explore De Quincey's gastrointestinal rhetoric. De Quincey hardly gives his reader an accurate account of opium's pharmacological effects, embedded as he is in

[12] In some respects, these entanglements of effect and side effect, drug and digestion, resonate with Brunonian systems of medicine, revealing the interaction of drug and body in the production of psychosomatic effects.

a Brunonian theory of his own physical distress.[13] Yet De Quincey's stomach becomes a site of significations that, like the metabolites of opiates, propagate effects across his body and his text. Because De Quincey makes his stomach crucial to his narrative of opium, the writing of opium-eating, like the pharmacokinetic effects of the drug, remaps and remakes the boundaries among action, effect and side effect, body and drug, centre and periphery, physiology and literature.

In drawing and redrawing these boundaries, De Quincey also entangles *cause* and *effect* in ways that rewrite the origins of opium-eating as one of its symptoms—gastrointestinal pain. In his 1821 *Confessions*, De Quincey claims that he first started taking opium regularly as a medicinal cure for these pains.

> It was not for the purpose of creating pleasure, but of mitigating pain in the severest degree, that I first began to use opium as an article of daily diet. In the twenty-eighth year of my age, a most painful affection of the stomach, which I had first experienced about ten years before, attacked me in great strength. This affection had originally been caused by extremities of hunger, suffered in my boyish days. (Lindop 2000, 13)

As Davenport-Hines observes, many opium-eaters, Samuel Taylor Coleridge included, first used opium as a cure for dyspepsia (2004, 71).[14] While De Quincey takes opium as a cure for what sounds like a bad case of indigestion, he also historicises his pain as a physical illness that emerges from the past of an autobiographical story he has yet to tell. His stomach pain works as an analepsis, recalling a memory that he is about to narrate and, in turn, creating the occasion for its narration. In this way, even before he becomes an opium-eater, De Quincey's status as an eater is already fraught with bodily and emotional pains that he embeds

[13] In his article, "De Quincey's Addiction", Morrison explores De Quincey's physiological accounts of addiction to suggest that De Quincey, in fact, gets opium wrong: modern pharmacology flatly contradicts De Quincey's protracted and painful depictions of withdrawal (Morrison 2009, 270).

[14] De Quincey acknowledges later in "The Pleasures of Opium" that the first time he tried opium was as an anodyne for a rheumatic toothache, suggested to him by a friend (Lindop 2000, 41). He draws a distinction between opium use and opium-eating, by suggesting the later describes the practice of taking opium daily rather than taking it as an occasional indulgence or remedy. For more on Coleridge's indigestion, see Gavin Budge's *Romanticism, Medicine, and the Supernatural: Transcendent Vision and Bodily Spectres, 1789–1852* (2012).

in his guts. These pains function as physiological memories, recalling experiences which occurred a decade before. The stomach pains recur, "attacking" "with great strength", as echoes from his "boyish days" when he was plagued by "extremities of hunger". In this sense, De Quincey claims that opium-eating occurs in the present to allay a memory of hunger from the past that asserts itself in his body as painful afflictions of his digestive organs.

However, these stomach pains are also potentially a physiological symptom of opiate use and withdrawal. In such descriptions, opium becomes a double cure: it allays a retrospective ill (extremities of hunger) that posits itself in the present as a potential side effect of use and withdrawal (intestinal cramping and chronic constipation) from that same cure. As Marc Redfield and Janet Farrell Brodie argue, in the circular logic of drug abuse, a drug helps to obliterate "memories of a pain [...] that would preexist it: the drug is at once pain and relief, poison and medicine; to be an addict is to live this circularity, with nothing at the origin but a sickness that strengthens itself in curing itself" (Brodie and Redfield 2002, 9). Like the Brunonian system of medicine, the temporality of drug use suggests that a cure is also potentially the cause of ill health. For De Quincey, laudanum cures the stomach pains and extremities of hunger that allegedly preexist it; yet it becomes increasingly clear in his *Confessions* that memories of hunger and their pains cannot be divorced from the laudanum he drinks. In this respect, there is a slippage among the extremities of hunger, stomach pain and opium-eating; they are not merely related phenomena, but rhetorical expressions of the same "sickness", that "strengthens itself in curing itself". Even as De Quincey's *Confessions* attempt to narrate a time before his addiction, his language, like his guts, are bound by rhetorical and physiological constrictions of laudanum as they operate on and across his body and text.

To examine the ways that the rhetorical and physiological constrictions of laudanum work on De Quincey's narration of hunger and illness, I focus on his descriptions of starvation and reflect on the ways that the illnesses he attributes to hunger are in fact produced rhetorically and narratively by the physiological effects of laudanum, becoming a "sickness" that "strengthens itself in curing itself". Early in his *Confessions*, he describes the period of starvation that produces his stomach pain: "I suffered [...] the physical anguish of hunger in various degrees of intensity; but as bitter, perhaps, as ever any human being has suffered who has survived it" (Lindop 2000, 21). The hunger reappears variously in his descriptions: A young child who De Quincey befriends during this time is "hunger-bitten", and she imagines or, "self-creates", ghosts as a result of the "real fleshly ills of the cold and [...] hunger"; De Quincey and the girl fall into "transient dozings at all hours", holding one another on the floor of the house of the lawyer Mr Brunell, in which he seeks refuge.[15] During periods of his "dog sleep", as De Quincey calls it, "a hideous sensation began to haunt" him, "a sort of twitching (I know not where, but apparently about the region of the stomach) which compelled me violently to throw out my feet for the sake of relieving it" (Lindop 2000, 22). De Quincey narrates a series of effects on the mind and body—pain, imagination, sleeping spells and "hideous sensations" in the gastrointestinal region—positing hunger in the same terms that will come to define the pains of opium. In this respect, the bodily and discursive marks of opium-eating shape the ways in which De Quincey remembers hunger itself, transforming the illness that caused his habit into a variation on the habit itself. In a sense, the rhetoric of hunger and starvation in De Quincey's text is another 'metabolite' of opium; that is, it is produced after opium has been consumed and becomes a potent *topoi* through which opium's effects are registered and understood. Eating and digestion in De Quincey's idiom then not only help us to understand the pharmacological importance of absorption, metabolisation, distribution and evacuation; they also figure the origins of opium abuse, rendered in the rhetoric of starvation, pain, hunger and intestinal cramping.

De Quincey, though, was not only an opium-eater; as Lockhart reported, he also drank alcohol in considerable amounts. Morrison even has suggested that De Quincey's dependence on alcohol rivalled his opium habit and produced more protracted and severe withdrawal pains. Physiologically, the symptoms of withdrawal from alcohol are far more dangerous than those from opium.[16] Morrison speculates that De Quincey "during the periods of his worst excesses was drinking the equivalent of between one and two pints of whiskey per day" (2009, 273). In this respect, not only was De Quincey subject to the intestinal cramping

[15] As he discusses in the 1856 *Confessions*, Brunell allows De Quincey to stay in the house, writing that "the attorney called himself, on most days of the week, Brunell, but occasionally [...] by the more common name of Brown" (Lindop 2000, 195).

[16] Morrison emphasises that while withdrawal from opiates usually lasts about a week and resembles the symptoms of the flu, alcohol withdrawal can last months and is potentially life-threatening (2009, 273).

that opiates produce, but these digestive problems were exacerbated by the alcohol in which the opium was dissolved as well as wine and spirits which were, by his own account, lifelong indulgences. Following his first episodes of starvation in the Confessions and the loss of Ann, the prostitute, on the streets of London, De Quincey leaves the city to seek financial help from an aristocratic friend (Lord D____) at Eton. When De Quincey arrives, his friend invites him to dine. However, when De Quincey sits down to eat, he has "no appetite" (although, he is eager to drink his friend's wine). He readily admits, "[o]n all occasions when I had an opportunity, I never failed to drink wine — which I worshipped then as I have since worshipped opium. I am convinced, however, that this indulgence in wine contributed to strengthen my malady: for the tone of my stomach was apparently quite sunk" (Lindop 2000, 34). Again, De Quincey remakes the boundaries of his opium use.[17] Opium and wine become one in the rhetoric of "worship", and their effects on his stomach assert themselves as a "malady" that both results from and produces desire for these sacred commodities.[18] The "tone of [De Quincey's] stomach" then becomes a signifier of chemical and rhetorical figurations of his "malady", which neither starts nor stops with the drug effects of opium, but entangles starvation, alcoholism, hunger and intestinal cramping with opium's other effects.

In this respect, De Quincey's rhetoric of 'opium-eating' is consistently bound up with other habits and symptoms, from starvation to alcoholism, and it then also designates various maladies as they constitute and reconstitute themselves in his intestinal tracts. It is no wonder his pharmacological explanations of opium addiction are inaccurate; for De Quincey 'opium-eating' designates a capacious space in which bodily processes of absorption, metabolisation, distribution and elimination confuse the contents of his stomach with his memories and his emotional life. In this sense, De Quincey's opium habit not only entangles one substance with another, effect with side effect, but generates a rhetoric of 'opium-eating' that reveals the imbrication of these physiological processes with narrative structure and memory. This is not to say that 'opium-eating'

[17] In another famous episode from De Quincey's *Confessions*, on the brink of death, he is revived by the prostitute Ann when she brings him a glass of wine (Lindop 2000, 26).

[18] For an extended analysis of the relationships between opium and De Quincey's religious rhetoric, see Charles Rzepka's monograph, *Sacramental Commodities: Gift, Text, and the Sublime in De Quincey*, published in 1995.

ultimately may signify anything in De Quincey's idiom, but rather that it captures the physiological, rhetorical and narrative enmeshments of cause and effect, mind and body—and even emotion.

While I have not focused primarily on affect or emotion in my accounts of opium's pharmacological effects, De Quincey gives us insight into the ways painkilling is never purely physical and sits uneasily between psychological and physical affliction.[19] There is no objective measure of pain, and, for this reason, its treatment with opiates or opioids remains difficult and controversial. Yet opium-eating in De Quincey's writing not only treats physical pain; it gives pleasures and pains that are both biological and emotional. In 1826, De Quincey confesses in a letter to his friend John Wilson that laudanum "has so much diminished" the "sensibility" "of my stomach […] that […] nothing ever stimulates my animal system into any pleasure" (Lindop 2000, 43; Morrison 2009, 249). In the nineteenth century, the word "sensibility" carries with it aesthetic and affective connotations, and De Quincey offers in this moment a counter-discourse to the moral aesthetic of taste in eighteenth and nineteenth century.[20] That is, the laudanum acts on his stomach as his stomach, digestive viscera and metabolic organs act on the laudanum, and in this ebb and flow of metabolic processes and metabolites, of opium-eating and eating opium, the ability of the stomach to operate as a discerning organ is greatly diminished. Presumably, the "sensibility" of his stomach refers to his appetite—his ability to take pleasure in eating. However, the "animal pleasures" to which he refers are bound up in the role digestive organs play in the regulation of mood (Wilson 2015, 100). That is, opium not only allays physical pain, but it also has ramifications for his emotional life. The metabolites of opium are not merely the by-products of the liver or kidneys as they pass into the bloodstream and the brain; they are also the moods and memories, diminished though they may be, as they are propagated across De Quincey's text. These moods and memories are themselves metabolites, which, like the opium his body

[19] For a Freudian interpretation of trauma and affect in De Quincey's texts, see Rei Terada's essay "Living a Ruined Life: De Quincey's Damage" in *Romanticism and Emotions* (2014).

[20] Denise Gigante's book, *Taste: A Literary History* (2005), traces the discourse of taste in British literary culture from Milton through the nineteenth century, arguing that it is a cultivated moral aesthetic that governs anxieties about consumption in the age of colonialism, urbanisation and industrialisation.

never stops 'eating', produces mounting psychosomatic and psychoactive effects. They are the product of a narrative and literary 'stomach' that consumes itself and, in consuming itself, produces the narrative account of his consumption.

Opium-Eating in the Twenty-First Century

There are many more examples of digestive tropes and stomach pains in De Quincey's writings, each of which can offer a deeper elaboration of 'opium-eating' as a rhetorical, narrative and physiological process. Rather than elaborating on them, I want to conclude by returning to the contemporary opioid epidemic in order to suggest that De Quincey's rhetoric of 'opium-eating' can reframe the moral and physiological assumptions that underlay addiction as a concept. The term epidemic implies that addiction is an infectious disease; Zieger, as I suggest above, traces its etymological roots to chattel slavery and reads the construct of addiction and moral attitudes towards addicts through a history of colonial violence (Zieger 2008, 5–8). Yet 'opium-eating' as I have traced it both in De Quincey's texts and through its pharmacokinetic mechanisms is not about either infection or enslavement; it is about metabolic processes and the emotional illnesses, maladies and symptoms of the digestive organs in response to these processes. Although De Quincey's habit operates within the circular logic of addiction that Brodie and Redfield describe—a "sickness" that "strengthens itself in curing itself"—as De Quincey illustrates, such a circular logic is also a fairly good definition of hunger. That is, hunger itself is never cured. Hunger, in being fulfilled, only begets, after a period of time, more hunger. To be sated, once and for all, would be death. So what De Quincey ultimately suggests is not that addiction infects or enslaves, but rather that it is closer to the mundane activities of daily consumption. It is significant, in this respect, that De Quincey takes opium as "an article of daily diet". In the Brunonian system of his time, opium, like food, is after all just one stimulant of organic excitation among many. That De Quincey's rhetoric of eating is as vexed as his rhetoric of opium use perhaps points less to the toxic effects of opiate abuse, than it does to the exigencies of consumption for which opiates and other habit-forming substances have always been a figure, implying, in some senses, that all forms of compulsory consumption fall along a continuum, ranging from eating to opioid dependence. In this respect, 'opium-eating' forces us to reckon with the 'eating' in opiate use and the ways in which opiate abuse itself might be understood as part of more general physiological and rhetorical narratives of hunger and emotion. Perhaps it is naïve to end an essay on the psychopharmacology of 'opium-eating' with an appeal to the figurative potential of De Quincey's rhetoric; however, it is in his figures that the ramifications of these pharmacological processes come to light and 'opium-eating' takes shape as a capacious process that traverses and remakes the boundaries of drug action, effect and side effect. The questions that De Quincey poses to our moment might be phrased like this: What narratives about opiate dependence might we be able to tell if we understood opium as one possible "article of daily diet" rather than an illegal drug? How might substance users be understood outside paradigms of addiction, infection and enslavement? If 'opium-eating' allowed De Quincey to narrate his memories, emotions, physiological afflictions and dependence as part of the same process, how might health-care workers, counsellors and law enforcement agents working against a drug 'epidemic' reframe opiate abuse as a complex interaction rather than the unilateral result of its ingestion? These questions may not have immediate answers, but they point the way towards an epistemic change that would allow addicts to become opium-eaters once more. This would not be a naïve return to the nineteenth century or to some pre-lapsarian fantasy of a world without physiological dependence. Rather, to allow addicts to become opium-eaters would be to acknowledge the enmeshment of medical, moral, literary and physiological discourses in the making and remaking of the boundaries between our bodies and our drugs on a continuum of hunger that will never be sated.

References

Beddoes, Thomas. 1795. Observations on the Character and Writings of John Brown. In *The Elements of Medicine of John Brown MD*, trans. J. Johnson Beddoes. Portsmouth: William and Daniel Treadwell.

Brodie, Janet Farrel, and Marc Redfield. 2002. Introduction. In *High Anxieties: Cultural Studies in Addiction*, ed. Janet Farrel Brodie and Marc Redfield. Berkeley, Los Angeles and London: University of California Press.

Brunton, Laurence, Bruce Chabner, and Bjorn Knollman. 2011. *Goodman and Gilman's the Pharmacological Basis of Therapeutics*. New York: McGraw Hill Medical.

Budge, Gavin. 2012. *Romanticism, Medicine, and the Supernatural: Transcendent Vision and Bodily Spectres, 1789–1852*. London: Palgrave Macmillan.

Burt, E.S. 2009. *Regard for the Other: Autothanatography in Rousseau, De Quincey, Baudelaire, and Wilde*. New York: Fordham University Press.

Clark, David L. 2003. We 'Other Prussians': Bodies and Pleasures in De Quincey and Late Kant. *European Romantic Review* 14: 261–87.

Davenport-Hines, Richard. 2004. *The Pursuit of Oblivion: A Global History of Narcotics*. New York: W. W. Norton & Company.

De Quincey, Thomas. 2009. *Confessions of an English-Opium Eater and Related Writings*, ed. Joel Faflak. Peterborough, ON: Broadview Editions.

Faflak, Joel. 2008. *Romantic Psychoanalysis: The Burden of the Mystery*. New York: State University of New York Press.

Gigante, Denise. 2005. *Taste: A Literary History*. New Haven: Yale University Press.

Grinnel, George. 2006. Thomas Beddoes and the Physiology of Romantic Medicine. *Studies in Romanticism* 45: 223–50.

Iseli, Markus. 2015. *Thomas De Quincey and the Cognitive Unconscious*. London: Palgrave Macmillan.

Lindop, Grevel. 2000. *The Works of Thomas De Quincey*, vol. 2. London: Pickering & Chatto.

Milligan, Barry. 1995. *Pleasures and Pains: Opium and the Orient in Nineteenth-Century British Culture*. Charlottesville: University of Virginia Press.

Morrison, Robert. 2009. *A Biography of Thomas De Quincey, the English Opium-Eater*. London: Phoenix Paperbacks.

———. 2011. De Quincey's Addiction. *Romanticism* 17 (3): 270–77.

O'Quinn, Daniel. 1997. The Gog and Magog of Hunnish Desolation: De Quincey, Kant and the Practice of Death. *Nineteenth-Century Contexts* 20: 261–86.

Quinones, Samuel. 2015. *Dreamland: The True Tale of America's Opiate Epidemic*. New York: Bloomsbury Press.

Rzepka, Charles. 1995. *Sacramental Commodities: Gift, Text, and the Sublime in De Quincey*. Amherst: University of Massachusetts Press.

Terada, Rei. 2014. Living a Ruined Life: De Quincey's Damage. In *Romanticism and Emotions*, ed. Joel Faflak and Richard Sha. New York: Cambridge University Press.

Vickers, Neil. 1997. Coleridge, Thomas Beddoes and Brunonian Medicine. *European Romantic Review* 8 (1): 47–94.

Wilson, Elizabeth. 2015. *Gut Feminism*. Durham: Duke University Press.

Youngquist, Paul. 1999. De Quincey's Crazy Body. *PMLA* 114 (3): 347–58.

Zieger, Susan. 2008. Introduction. In *Inventing the Addict: Drugs, Race, and Sexuality in Nineteenth-Century British and American Literature*, ed. Susan Zieger. Amherst: University of Massachusetts Press.

CHAPTER 7

A Posthumanist Approach to Agency in De Quincey's *Confessions*

Anna Rowntree

"Not the opium-eater, but the opium, is the true hero of the tale; and the legitimate centre on which the interest revolves", writes Thomas De Quincey in the second decade of the nineteenth century (2013 [1821], 77). These words, found in *Confessions of an English Opium-Eater*, suggest the power of opium to direct the narrative of a person's life, and they problematise the idea of agency as a uniquely human quality. De Quincey was a man who oriented not only his writing around the centrifugal force of opium, but whose daily reality was concentrated upon the drug, its procurement, preparation and consumption. He was what we would today call an addict, although medical science of the nineteenth century was yet to define habitual drug use in this way. Indeed, how medical men throughout the nineteenth century started to account for the seemingly problematic effects of opium use (whilst continuing to recommend it) tells us more about the biases and anxieties of the day than it does about the phenomena of addicted behaviour or the subjective

A. Rowntree (✉)
Berlin, Germany

N. Roxburgh and J. S. Henke (eds.), *Psychopharmacology in British Literature and Culture, 1780–1900*, Palgrave Studies in Literature, Science and Medicine,
https://doi.org/10.1007/978-3-030-53598-8_7

experience of drug dependency. De Quincey, on the other hand, writes from inside the drugged experience, offering insights that are challenging not only to nineteenth-century medical practitioners and moralists but also to our own normative idea of the human. The following analysis, therefore, aims at a contextualised reading of De Quincey's *Confessions* by taking into account a vital materialist understanding of the substance. It does so in order to point towards the possibility of a posthumanist understanding of opium as agentic, in contrast to Victorian medicine's humanist approach to addiction as an illness to be cured by raw willpower. This chapter thus shifts attention away from addiction and focuses instead on what was known about the drug's psychopharmacological effects.

We cannot talk about habitual drug use as it was understood then or addiction as it is understood now without facing the knotty and painful fact that addicts "appear to grow estranged from their own behaviour" (Weinberg 2013, 175) and that "loss of self-control" forms "the core criterion of addiction" (173). How can we account for an individual's apparent loss of autonomy and rational self-determinacy? Both in the common imagination and in medical discourses, the loss of self-control exhibited by the human when under the influence of physiologically and physically affecting substances is the primary feature distinguishing recreational, allegedly harmless, drug use and pathological use (Weinberg 2013, 173). It is also a theme richly explored in *Confessions*. Despite the early sublime effects of opium, by the latter half of *Confessions* we see De Quincey exhibiting mental and physical torpor, as though chained to his bed as "powerless as an infant" (2013 [1821], 66), victim of a drug he describes as a "tyrant" (77). The language of victimisation, oppression and powerlessness is provocative, leading even modern-day commentators such as Richard Davenport-Hines to write derisively about what he sees as De Quincey's "puerile helplessness" (2004, 68).

It is certainly tempting to understand the construction of a victim position in De Quincey's writing as "a strategy of disavowal that allows him the luxury of denying responsibility for his addiction" (Schmitt 2002, 73). And, in a similar vein, one might also want to apply the same logic to all those claiming a loss of self-determinacy, but doing so risks avoiding an important and fruitful challenge. Taking seriously De Quincey's impression that his ability to act is compromised by another agent forces us to rethink our understanding of responsibly, agency and control. We are thus encouraged to contextualise *Confessions* within other arguments about the nature of human life, observing how De Quincey begins to think through the way drug dependency undermines the notion of an autonomous individual at a time when medical men were struggling to understand the relationship between drugs and the human body. This is a project with deeply modern resonances, not only because current conceptions of addiction continue to express a strong humanistic bias centred around human autonomy and agency but also because a posthumanist theoretical framework offers a uniquely satisfying way to understand addiction generally and De Quincey's proto-vital materialism specifically.

Darin Weinberg, in his *Posthumanism, Addiction and the Loss of Self-Control*, describes what he sees as the failure of both biomedical and social scientific approaches to explain the relationship between loss of self-control and drug use. He too believes that posthumanism offers a uniquely satisfying way of describing the drug/human relationship, one freed from restrictive beliefs about human autonomy, agency and self-determinacy (Weinberg 2013, 173). It does so because under a posthumanist lens it becomes impossible to differentiate the human from the physical contexts in which it is engaged. The border between inside and outside blurs, and the body looks porous, mutable and shifty. It becomes an interactive conglomerate thing in dynamic interchange with other things. The drug, a thing or body in its own right, may begin as external to the human body, but soon after ingestion it becomes internalised and entangled. The human is not only not independent of the material world but is also not exclusively powerful. Humanist phantasies about the self-defining, rational person in dominion over the material world are replaced by an understanding of the human as shaping and being shaped by nonhuman forces of all kinds.

In order to enable a posthumanist reading of drugs and their effects, it is necessary to first consider the agentic nature of the material world as such. In *Vibrant Matter: A Political Ecology of Things*, Jane Bennet provides a detailed description of the intrinsic vitality of all matter and assemblages of matter. The concept of *assemblages* is one borrowed from Gilles Deleuze and Félix Guattari and employed by Bennet to argue that the human is a gathering of lively matter (skin cells, intestinal bacteria and so on) and therefore an example of vibrant materiality no different from other gatherings of lively matter (Bennet 2010, 23). Seen in this way, the individual is never quite as individual as assumed, and the human is always already in dynamic interchange with the nonhuman of which it is formed. Moreover, Bennett gives voice to thing-power in order to shift the focus away from an exclusively human understanding of agency and autonomy. She illustrates how food, stem cells, electricity grids and even metal can demonstrate their ability to "act as quasi agents or forces with trajectories, propensities or tendencies of their own" (Bennet 2010, viii). The ability of allegedly inert matter to affect other bodies is the starting point for arguing that humans are not the only political, cultural or social actants. She uses Nietzsche to describe edible matter "as a powerful agent, as stuff that modifies the human matter with which it comes into contact" (Bennet 2010, 44). Fats, for example, illustrate how "human wills, habits and ideas" can "weaken or enhance" under the power of edible matter (43). Faced with a packet of crisps, Bennett finds herself entering "an assemblage in which the 'I' is not necessarily the most decisive operator" (40). Training oneself to see in a way which pays attention to the interactions between "human" and "world", and the ill-definable border between the two, encourages a way of understanding the human which is fundamentally challenging to a humanistic idea of the rational, self-governed and self-determined human. It is no longer possible to assume that agency and the power to exert influence, govern or modify are exclusively human traits. Instead we become aware of the many nonhuman factors displaying the power to effect change and directionality and must descend from atop the traditional Christian or even humanist hierarchy, which sees us enjoying dominion over the animals, plant life and inanimate things of the world, in order to level the playing field between bodies of all types.

For an example of how a traditional hierarchy informs a modern understanding of drugs and addiction, we need look no further than John Brown's influential *Elements of Medicine*, published between 1780 and 1795. The system of medicine proposed in Brown's work, influential in the first decades of the nineteenth century and known as Brunonianism, relies on an understanding of disease as an imbalance in excitability. The body is either over- or under-stimulated, the job of curative drugs is therefore to correct this imbalance through stimulation or sedation. Brown theorises a relationship between the human and the world, which acknowledges the fluctuating nature of the human body and the power of external factors to affect deep internal change. There are intriguing hints of the kind of affect theory that modern posthumanism relies upon. What is also noticeable through his careful cataloguing of types and strength of stimulation, however, is Brown's belief in the measurable nature of drug effects and human energy levels. He writes: "Exciting powers may be referred to two classes. 1. External, as heat, food, wine, poisons, contagions, secreted fluids, and air. 2. Internal, as the functions of the body itself, muscular exertion, thinking, emotions and passions" (Brown 1795[1790], 12). External substances such as wine are seen to affect the internal workings of the human individual, but only insofar as they stimulate or sedate. Despite several hints about a fluctuating human body, therefore, there is no suggestion that the clearly defined border between what is internal or external to the human subject is questioned. Substances are merely objects of use taken into the body but not integral to the human's sense of himself or herself.

In the Brunonian system, medicine is a purely biological matter and medicinal drugs when correctly applied according to their stimulating or sedative effects do nothing more than return the individual to health, as here with opium: "Opium is not a sedative; on the contrary, as it is the most powerful of all the agents that support life, and that restore health, and a truly blessed remedy, to the divine virtue of which the lives of so many mortals have been owing" (Brown 1795[1790], 243). By describing the drug's power as divine, Brown resorts to that traditional ontological hierarchy which sees humans as superior to all matter, governed over only by God. Power as great as opium's cannot be seen as intrinsic to the drug itself without admitting a flatter ontology in which humans and matter are levelled and the human begins to look vulnerable. This acknowledgement of human vulnerability is something that Brown's system—which focuses on opium in particular—does not allow.

Vulnerable is indeed what human beings are. Not only to the unwanted side effects of drugs (which Brown goes some way towards addressing) but also to the unpredictable social, cultural and individual effects of the drug/human assemblage. When we turn to the question of habitual consumption, the drug becomes more than a cure for disease, proving integral to a person's mode of being in the world. It is only when dealing with the strange problem of habitual consumption—what later comes to be called *addiction*—that nineteenth-century medical practitioners begin to move beyond an understanding of drugs as only superficially effecting. As Bennett argues, humans have traditionally imagined themselves as not only "organic, unique, and ensouled" but also as occupying "the top of the ontological hierarchy, in a position *superior* to everything else" (Bennet 2010, 87). With comfortable assumptions about our place in the world questioned, the consumer of substance we now call the *addict* becomes not just a social irritant but also a fundamentally threatening example of human powerlessness. This is why it is significant that, out of this context of uncertainty, new approaches to habitual drug use emerge, exemplifying the role science had in mitigating the spiritual and social uncertainties of the age.

It is also interesting to note the significant influence of Brown's system on writers and thinkers contemporary to De Quincey. In 1794–1796, Erasmus Darwin, grandfather to Charles Darwin, published *Zoonomia, or the Laws of Organic Life*, a part medical, part philosophical treatise on the nature of life that clearly foreshadows his grandson's theory of evolution by suggesting the ability of all life to adapt and develop certain characteristics based on needs, desires and aversions. More than this, however, he argues that all animals, from humans to amphibians, have "seem to have been formed by the original living filament" (1818, 397), suggesting if not exactly a common ancestor then a common liveliness to all being. In an attempt to unite medical knowledge under this one grand principle, he describes the four kinds of motion common to all life and follows Brown in seeing stimulation as an essential factor. Interestingly, in illustrating the animated nature of organic life he defines himself against anyone "attempting to explain the laws of life by those of mechanism and chemistry" saying "they considered the body as an hydraulic machine [...] forgetting that animation was its essential characteristic" (vii). Darwin therefore attempts to unite medical knowledge under one grand principle (which he describes in physiological detail) whilst at the same time pointing towards a "living principle, or spirit of animation, which resides throughout the body, without being cognizable to our senses, except by its effects" (5). We might remind ourselves of Bennett's crisps here—and that mysterious undercover liveliness only made obvious to us by the mysterious stretch towards the packet. Bennett of course sees this liveliness as a property of diverse assemblages, whereas for Darwin "[t]he ultimate cause of such motions of life can only be hazily sketched" (Packham 2014, 151).

We therefore see Darwin posed between an empirical understanding of the body and a materialist's appreciation for the mysterious, unknowable life of matter. His is a materiality which is inclusive, mutable, environmentally responsive and affective. What Darwin does not do is apply his understanding of animation to include non-organic matter, although we might find him playing with the idea in some of his poetry.[1] What he does

[1] See essay in this volume by C. A. Vaughn Cross.

do in *Zoonomia* is consider plant life, seeing the same principles of animation, and the ability to self-direct in both plants and humans. He even goes as far as to wonder if plants have an emotional and sentient ability. Whilst anthropomorphising plants, he reveals an anthropocentric perspective that cannot be called posthuman, but it is nonetheless intriguing to see ideas such as love and consciousness being applied beyond the human.

Over the course of the nineteenth century, concepts of addiction gained currency in large part due to medicine's desire to define and sanction human behaviour and social norms. One sees this in many of the early nineteenth-century discourses on alcohol, which, to a large degree, reflect a similar logic as those on opium. Thomas Trotter's 1804 *An Essay, Medical, Philosophical, and Chemical on Drunkenness, and Its Effects on the Human Body* provides an early example of habitual substance use entering the medical consciousness. Although he does not deal directly with opium, Trotter writes that "other narcotics" such as opium produce "nearly the same phenomena" and "almost the same disease" as habitual alcohol consumption, already pointing to a commonality in the power of drugs which exceeds a substance's specific chemical effects. Defining habitual substance use as a disease is for Trotter a way to move beyond religious and moral commentary in order to look more widely at the physical, social, personal and emotional causes underlying habitual drinking (1804, 3–4). However, in the years following Trotter's effort, it becomes clear that *disease* is by no means a neutral term and that medicine and morality are inescapably linked when it comes to the subject of habitual drug use.

Louise Foxcroft, in her book *The Making of the Addict*, persuasively points to a Victorian faith in the medical sciences in "an increasingly secular environment" and argues that the "cult of 'scientism', invested with credence by evolutionary theory, gave a foundation to an approach to the human condition which was observable and 'factual'" (2016, 168). Foxcroft's argument is that faith in the sciences, in replacing a faith in God, helped "sanction the dominant moral code" by describing drug addicts as unnatural and diseased (168–69). She argues that the Victorian belief in medicine to cure those aspects of behaviour deemed diseased can be seen as an attempt to cleanse society (168). By describing the individual addict as other, deviant or diseased, it is possible to ignore the challenge of the drugged human to a normative understanding of ourselves. At the same time, Foxcroft describes how by the 1880s the medical profession was engaged in a "'war' on opium which was seen to be a dangerous poison, an agent of vice, and often fatally destructive in the wrong, non-medical hands" (167). This move was not designed to humble mankind by pointing to opium's thing-power but rather an attempt by the medical profession to "monopolise control of opium use" (167). Much as with Brown, the medical man is seen as uniquely empowered to handle opium. Gone, however, is Brown's spiritual terminology, which is replaced by a significant note of anxiety. Opium's known drug effects are seen as both physically and culturally dangerous, and there is a rush to establish a new scientific social order to account for and control opium and opium addiction.

This move, compelled by the professionalisation of medicine, is one based on empirical explanations. For, as Foxcroft suggests, the need to understand and even cure habitual drug use has historically led to and resulted from a difficult mind/body split in the understanding of addiction: "[T]here existed a clear and unbridgeable distinction between science and the mind", she writes, with medical knowledge unable to account for or measure the kind of "creative, mystical experience as described by De Quincey *et al*" (Foxcroft 2016, 166). So whilst the sciences set about understanding the biological factors involved in addiction, the subjective, emotional or even spiritual aspects of drug consumption were left to the writers and philosophers. According to Weinberg, it is a fault line still visible in addiction studies today, with biomedical sciences and the social sciences inhabiting different sides of that divide. His effort to theorise addiction through the posthuman is an attempt to correct what he sees as this "overwhelming tendency to conceptualise human biology and human social life dichotomously as two, and only two, wholly discrete and independently integrated ontological domains" (Weinberg 2013, 173). Reading opium and De Quincey through a notion of materialism prevalent in Romanticism (Goldstein 2017, 7) goes a long way towards correcting this tendency, enabling us to pay attention to the non-empirical effects of opium consumption whilst still keeping in view developments in medical science. Doing so also allows us to reject a strict Victorian understanding of matter in favour of proto-vital materialism more relevant to De Quincey himself.

Taking into account the likely influence of commentators such as Brown, Darwin and Trotter, it becomes apparent that by the time De Quincey wrote *Confessions*, the nature of human and material life was already being creatively and publicly considered. In the light of Darwin's plant theory in particular, De Quincey's claim that opium (itself a plant derivative) is the hero of *Confessions* begins to look less eccentric. I would suggest, however, that beyond demonstrating the thing-power or vitality of opium, what we get in *Confessions* is a fuller de-centring of the human than we observe, for example, in Darwin. This is accompanied by an interrogation of humanistic assumptions about autonomy, control, rationality and knowability that hints at a more modern posthumanism.

At first, it may seem counterintuitive to read autobiography (that most anthropocentric of genres) through a posthumanist lens. The life story is in many ways a triumph of human agency. It seems to stand as a testament to language, rationality and individual endeavour. However, for those paying attention, there is something strangely nonhuman about autobiography and the autobiographical subject, which undermines a simple humanistic narrative. Bennett says, "The sentences of this book also emerged from the confederate agency of many striving macro-and microactants" (Bennet 2010, 23). By this she means that everything from the birdsong outside her window to her glasses was involved in the production of her text. Whilst this may be true, there is also something else about the autobiographical subject that makes it more obviously a *thing* than we usually acknowledge the human to be. I am talking not only about the materiality of the book (ink and paper) but about the way in which we objectify the human through the act of writing, lending it something of the character of inert matter. The subject of a life story is a thing which walks and talks like a human and yet it is handled and observed like a pebble. We describe and define. The author, in this case De Quincey, stands apart from this thing that both is and is not himself. At the same time, to tell one's life story is to describe a body in context, enabling readers to watch a character emerge from an infinity of colliding influences and to witness the development of a subject.

This treatment of the human is especially explicit in *Confessions*, where De Quincey plays with the border between the explicable and the lively. The text opens with the words "To the reader-" and a careful apology for the impolite confessions that will follow. "Nothing", De Quincey writes, "is more revolting to the English feelings, than the spectacle of a human being obtruding on our notice his moral ulcers or scars, and tearing away that 'decent drapery', which time, or indulgence to human frailty, may have drawn over them" (2013 [1821], 3). He places in front of the audience an item of grotesque interest, inviting readers to view the subject as if a thing. The premise of *Confessions* is that De Quincey will unflinchingly detail and observe his relationship with opium, the effects and affects of it on his life and body. It is in many ways a scientific endeavour, and on many occasions De Quincey suggests that the medical man could learn much from the account. It is also, however, a project which pays attention to the emotional, erratic and incomprehensible symptoms of drug use. In this way, De Quincey positions his understanding of opium against a medical understanding developed without direct phenomenological experience of the drug. He writes that even "those who have written expressly on the materia medica, make it evident, from the horror they express of it [opium], that their experimental knowledge of its action is none at all" (2013 [1821], 43). De Quincey therefore writes from "inside" the drug experience, simultaneously offering up his experiences as evidence useful to a scientific understanding of the drug. His is an effort to introduce empirical science to the lively and personal. Literature is not seen as irrelevant to medical science; instead, spiritual or moral questions arising from the consumption of the drug are seen as appropriate to a full scientific understanding of its functions. We might again think here about the mind/body split present in Victorian and modern addiction studies and compare them to a vital materialist's effort to merge the biological and the social.

Interestingly, in the opening lines, morality, that allegedly distinct human characteristic, is not figured as divine or God-given. Instead De Quincey writes about the "human [...] obtruding on our notice his moral ulcers or scars" (2013 [1821], 3). Morality is therefore something prone to disease and scarring. Polite convention may have drawn a "decent drapery" (3) over human fragility, but the true human, stripped bare of social etiquette, is presented as a base bodily thing. Morality is not a stable divine quality but something dynamic, corruptible and physically located. This is the first indication of De Quincey's nuanced mind/body relationship. Moral faculty, which would traditionally be attributed to the mental, rational aspect of the human (thus justifying a view of the human as superior to the purely corporeal animal), is for De Quincey a material matter. We see De Quincey beginning to break down mind/body dualism as he attempts to understand the spiritual, intellectual and moral implications that bodily entanglement with opium entails.

We should be careful not to suggest that De Quincey fully breaks down the mind/body split, however. Instead, the ways in which he describes mind and body appear shifting and unresolved. To a large degree, it is the assertive presence of the body which facilitates this complexification of Cartesian dualism. Throughout *Confessions*, the body is the site of struggle. Bodily pain promotes the use of opium, and it is the body that towards the end of *Confessions* is chained down and unresponsive (De Quincey 2013 [1821], 66). So whilst there is none of the mind/body oneness we might expect to find in today's posthumanism, the central role played by the body is nonetheless significant.

By repetitively drawing attention to the physical aspects of human experience, De Quincey undermines a belief in the pre-eminence of the mental faculties to direct the trajectory of human life. Drawing attention to the role of physical pain and other external factors (such as bereavement and financial hardship) in directing the course of a life allows De Quincey to trace lines of affect and indirect causation from his early teenage years into the later years of habitual opium use, making nonsense out of the question: "How came any reasonable being t o subject himself to such a yoke of misery?" (De Quincey 2013 [1821], 6). The reasonable human is far too faulty a notion to account for De Quincey's relationship with opium. For De Quincey, opium use is not the outcome of a rational decision (see Fay 2010, 293). The trajectory of his life is seen to emerge from a milieu of diverse factors over which De Quincey feels himself to have little control. Physical and emotional pain and the easy availability of opium each play a part. What this means is that the power to effect change is not tied to human will or rational action. Circumstance created by human and nonhuman factors, including the powerful urge of a body in pain and the profound effects of the drug itself, all have a role in creating De Quincey the opium-eater. Reason is an unavailable luxury.

This inevitably affects how De Quincey thinks about responsibility and morality. Following his careful justification for writing *Confessions*, De Quincey seeks to assuage himself of any guilt. The guilt of self-indulgence is called into doubt on the grounds that the relief of pain motivated much of his opium consumption. Later, describing his sorry condition in 1813, De Quincey writes that both a "distress of mind connected with a very melancholy event" and an attack of "the most appalling irritation of the stomach", resulted in an increased consumption of opium. He is also at pains to justify himself: "I postulate— that, at the time I began to take opium daily, I could not have done otherwise" (De Quincey 2013 [1821], 53). Clearly the human intellect is not the only agent exerting power. The consumption of opium on a daily basis comes about as a result of complex, physical, emotional and contextual factors over which De Quincey feels himself to have no control. We might imagine how his contemporaries read his apparent loss of agency. For those medical practitioners claiming complete knowledge of the drug's effects and their ability to safely prescribe it, De Quincey's condition is most safely described in terms of pathology and moral failure.

De Quincey, predicting this, levels his argument against those he imagines moralising against him: "An inhuman moralist I can no more endure in my nervous state than opium that has not been boiled" (2013 [1821], 53). The humorous pulling-together of the moralist and the opium suggests the similar and dangerous nature of both. Both the moralist and the drug are characters exerting potential influence over De Quincey. The human actant is not only "inhuman"; he is not any more powerful than the un-boiled opium. The moralist's "hard words" are laughably puny in the face of the energetically exhausted De Quincey bent on opium consumption (54). We might think here about Brown's influential assertion that opium works as a stimulate for those suffering energetic depletion. Whereas Brown assumed that after a dose of opium, the human would return to self-regulating health, De Quincey demonstrates how opium becomes an embedded part of his body's system necessary for normal function.

De Quincey goes on to highlight the gap between moral ambition and bodily reality: "The opium-eater loses none of his moral sensibilities or aspirations [...] but his intellectual apprehension of what is possible infinitely outruns his power, not of execution only, but even of power to attempt" (2013 [1821], 66). Here he understands the mind and body as separate. Moral sensibility and ambition are located intellectually, but these do not translate into physical action. In this equation, the body is more powerful than the intellect. It is the body which entangles De Quincey with the drug and in so doing acts beyond the remit of the intellectually located and morally intact self.

Attuning ourselves to the lively nature of assemblages of matter might not unknot the problem of moral responsibility, but it forces us to face a conundrum at the heart of what it means to be human. When De Quincey calls into question easy moral assumptions about the nature of drug use and when he insists that *Confessions* is less about himself and more about the power of opium, he is calling attention not only to "the marvellous agency of opium" (2013 [1821], 77) but also to the fallacious belief in human prowess and control. De Quincey is a man of prestigious intellect, a scholar and philosopher, and yet *Confessions* is a tribute to forces working beyond and even against the will of the human mind.

Confessions is also rich with moments where human comprehension, and not just human ambition, is exceeded. Whilst much of *Confessions* conforms to an analytical tone suited to the scientifically unflinching nature of the project, at times—notably, when in a state of reverie but also during the dream episodes for which *Confessions* is probably most famous—De Quincey opens up a space for the incomprehensible and indescribable. Again we are forced to acknowledge a power which is wholly beyond the scope of the humanistic language-wielding person. In the introduction to the dream episodes, De Quincey describes a melancholy "incommunicable by words" and "proportions so vast as the bodily eye is not fitted to receive" (2013 [1821], 68), suggesting not only an exceeding of language but an exceeding of human ability. He also describes the "sympathy" which arise between "the waking and dreaming states of the brain" (67). Consciously and voluntarily evoked thoughts are seen to translate into dreams. At first, thoughts turn into faint hallucinatory "phantoms" and then, "by a process apparently no less inevitable […][,] they were drawn out by the fierce chemistry of my dreams, into insufferable splendour that fretted my heart" (67). This move from conscious thought into hallucination and then uncontrollable dream reality is experienced by De Quincey as inevitable and independent of his will or understanding.

On the one hand, there is the observing, conscious self and, on the other, a part of the mind which is outrunning the limited scope of consciousness and control. We might call this latter part the unconscious and see De Quincey prefiguring Freudian psychoanalysis (see Iseli 2019, 325). Or we might look again at Darwin and his influential ideas about associated trains of thought, in which a sensation automatically calls up other memories, sensations or thoughts associated through habit with the original stimulus. Darwin sees these trains of association as mitigated during the waking hours by the presence of the material world impressing itself upon us, whereas at night, disconnected from other stimuli, we relive imagined sensations much more vividly. There is a "reiteration" of the original lived sensation, suggesting not only that the body uninhibited by the conscious mind expresses a life of its own but also that the body has a kind of emotional memory, shaped by the sensations earlier visited upon it.

Compellingly, for Darwin and De Quincey, this unconscious aspect is understood as biological, a function of the body. In De Quincey it is "the brain", the "eye" and the "fierce chemistry" (2013 [1821], 67) of his dreams that facilitate the move from thought to hallucination and then dream. It is as though the body is able to make an affectual connection that moves beyond the conscious rational self, opening the subject up to new incomprehensible experiences. In this way, the body (eye, brain, chemistry) is seen as mysterious, its processes outside of the control of self. De Quincey is unable to stem the rush from thought to sensory perception (hallucination) and virtual reality (dream).

In *The Posthuman*, Rosi Braidotti describes the mutation of "human embodiment and subjectivity", which she sees as pertinent to our current moment, as "throwing open the doors of our collective perception, forcing us to hear the roar of cosmic energy that lies on the other side of silence and to stretch the measure of what has become possible" (Braidotti 2013, 196). I understand this as exposure to the otherness which exists beyond our concepts of self. We are reminded "that we are, after all, such stuff as dreams are made of and that the new possibilities are immense" (197). In the case of De Quincey, however, this process of movement beyond the conscious and controlled self and the exposure to incomprehensible otherness is less enthusiastically embraced. Nonetheless, I see De Quincey pointing to forces beyond the scope of rational control, pre-empting a posthuman reconsideration of the rational, self-controlled human. Even more than this, it is in the dream episodes where we see the drug and the human most palpably in assemblage. The drug is more than inert matter to be refined and traded; it is, in combination with the human, a thing with a life in excess of empirical understanding. The drugged body/embodied drug expresses a propensity and unpredictable directionality of its own.

Opium is not only a product, an incidental catalyst of bodily and social change; it is a thing we cannot hope to measure, a thing so other-worldly that De Quincey cannot quite believe that the druggist who first sells him it is even mortal (2013 [1821], 38). In simply framing *Confessions* around the dual possibilities of pleasure and pain, De Quincey is undermining any narrow understanding of opium. It is an agent of divine revelation and horrific hallucination. Opium is both the central protagonist and a thing that slips through the fingers, never to be pinned down. Put simply, the De Quincey-opium assemblage reveals the entangled nature of human life and the lively nature of the nonhuman. De Quincey undermines a purely empirical effort to understand the drug and habitual drug use. He seems to talk directly to a medical profession intent on defining the nature, character and affordances of the drug, instead revealing how opium continues to escape capture, always exhibiting a life in excess of objective understanding.

Efforts to describe and delimit the drug effects of opium are seen to be doomed, for to do so would be also to define and limit ourselves, to draw a clear unbridgeable line around what is human and what is not. What the addict knows all too well is that separating self and drug is not a matter of arbitrary identity categories. The border between self and world is in constant negotiation. Who or what exerts power, our experience of embodiment, what we understand as external to ourselves and what has become incorporated into our being: these things are in constant flux. As Bennett argues, we both are and are not what we eat. De Quincey both is and is not opium.

Cleary i n this field of interactive elements, it is not possible to definitively say what the human is and is not; the location of a personal identity is difficult. Davenport-Hines sees De Quincey failing to "appease mid-Victorian moralists" on the grounds that his sense of identity "was often unstable, improvised, disintegrating and discontinuous". He says: "Habitual users of narcotics seldom fit into the bourgeois sense of human identity as a serious business, stable, abiding and continuous, requiring the assertion of one true cohesive inner self as proof of health and good citizenry" (Davenport-Hines 2004, 70). Arguably, it is in part this instability which stimulates so much anxiety about the drug user both in the nineteenth century and today. The drugged human reveals the mutable and plastic nature of human identity and the powerful nature of nonhuman matter.

Instead of anaesthetising ourselves to this anxiety by othering the addict, might we not use it to enable a more general re-examination of the human/nonhuman relationship? Seeing addiction as moral degeneracy and/or disease is a convenient way to distance ourselves from the agonising realities of the addict, preventing us from responding constructively to the challenge the addict poses to humanistic ideas. By contrast, seeing addiction as an extreme demonstration of our material vitality prevents us from dehumanising the addict and encourages us to think more carefully about how we all engage with the stuff of the world. Addiction studies, like any field of study, tends to compartmentalise the subject, so whilst biologists and neuroscientists are examining the effects of hormonal imbalance and chemical interference on the brain, and psychologists are piecing together the fragments of a half-forgotten childhood, humanities scholars are looking to culture, history and language.

Although all of these approaches have their benefits, the man or woman on the street remains a conglomerate entity of all these factors. Posthumanism reminds us that the human is shifty and that clear lines of identity are impossible to draw. This, in turn, helps us to avoid the habits and biases of our academic discipline. Bennett's notion of vital materiality is additionally helpful in that it directly addresses and rethinks human responsibility, a concept intrinsic to the study of habitual drug use. She writes: "Autonomy and strong responsibility seem to me to be empirically false, and thus their invocation seems tinged with injustice. In emphasising the ensemble nature of action and the interactions between persons and things, a theory of vibrant matter presents individuals as simply incapable of bearing *full* responsibility for their effects" (Bennet 2010, 37). Her argument is that avoiding "the blame game" expands the places we might look for the sources of harmful effects. Taking a vital materialist's approach to habitual drug use forces us to look beyond the individual, even beyond culture, to pay attention to the multiple and interactive forces at play, both human and nonhuman.

To be alive today is to be confronted by what it means to be a human in a globalised technology-mediated world where stimuli overload is a constant risk. De Quincey is just one example of a person beginning to think through the consequences of being a body relationally entangled with the world. Where does power lie? How can we assume responsibility? What are our moral obligations and capabilities? These are questions pertinent to posthumanism, De Quincey, psychopharmacology and addiction studies. The hope of new materialists like Bennett is that facing these questions will enable us to engage more ecologically and ethically with the world. At the end of *Confessions*, De Quincey claims that the "moral of the narrative is addressed to the opium-eater" (2013 [1821], 78). If he can teach the opium-eater "to fear and tremble", then he has achieved enough (78). I suggest that he achieves much more than that. By demonstrating the power of the drug, De Quincey not only teaches the opium-eater to tremble, but the undrugged reader, too. It is hard to read *Confessions* and still believe in the dominion of man over all else. Witnessing the thing-power of opium is a humbling experience. Today, it is arguably humility which we are all most sorely in need of.

References

Bennett, Jane. 2010. *Vibrant Matter, A Political Ecology of Things*. Durham: Duke University Press.

Braidotti, Rosie. 2013. *The Posthuman*. Cambridge: Polity Press.

Brown, John. 1795 [1780]. *The Element of Medicine of John Brown M.D*, trans. J. Johnson Beddoes. Portsmouth: William and Daniel Treadwell.

Darwin, Erasmus. 1818 [1801]. *Zoonomia, or the Laws of Organic Life*, 2 vols. Philadelphia: Edward Earle.

Davenport-Hines, Richard. 2004. *The Pursuit of Oblivion: A Global History of Narcotics*. New York: W. W. Norton.

De Quincey, Thomas. 2013 [1821]. *Confessions of an English Opium-Eater*, ed. Robert Morrison. Oxford: Oxford University Press.

Fay, Elizabeth. 2010. Hallucinogesis: Thomas De Quincey's Mind Trips. *Studies in Romanticism* 49 (2): 293–312.

Foxcroft, Louise. 2016. *The Making of Addiction: The 'Use and Abuse' of Opium in Nineteenth Century Britain*. London and New York: Routledge.

Goldstein, Amanda Jo. 2017. *Sweet Science: Romantic Materialism and the New Logics of Life*. Chicago and London: University of Chicago Press.

Iseli, Markus. 2019. Thomas De Quincey and the Cognitive Prospects of the Unconscious. *European Romantic Review* 24 (3): 325–33.

Packham, Catherine. 2014. *Eighteen-Century Vitalism*. London: Palgrave Macmillan.

Schmitt, Cannon. 2002. Narrating National Addictions. In *High Anxieties, Cultural Studies in Addiction*, ed. Janet Farrell Brodie and Marc Redfield. Berkeley, Los Angeles, and London: University of California Press.

Trotter, Thomas. 1804. *An Essay, Medical, Philosophical and Chemical, on Drunkeness and Its Effects on the Human Body*. London: Longman, Hurst, Rees, and Orme.

Weinberg, Darin. 2013. Post-humanism, Addiction and the Loss of Self-control: Reflections on the Missing Core in Addiction Science. *International Journal of Drug Policy* 24: 173–81.

PART III

The Cultural Politics of Known Drug Effects

CHAPTER 8

Reading De Quinceyan Rhetoric Against the Grain: An Actor-Network-Theory Approach

Anuj Gupta

The rhetorical strategies that authors use to structure texts reflect the aesthetic temperaments as well as the epistemic categories that are prominently used to frame discourses in the unique historical moment that they belong to. Analysing De Quinceyan rhetoric in his *Confessions of an English Opium-Eater* (1821) should thus tell us something about the ways in which drugs and their actions, especially opium, were understood, contextualised and represented in the collective psyche of nineteenth-century Britain. How does this text, considered to be canonical within the cultural representation of psychotropic substances, frame the drug effects of opium? What was its relationship with the prevalent contemporary medical discourses of its times and how were these discourses, in turn, conditioned by utilitarian epistemic frameworks? What ways of

A. Gupta (✉)
Rhetoric, Composition & the Teaching of English (RCTE) Program, Department of English, University of Arizona, Tucson, AZ, USA

N. Roxburgh and J. S. Henke (eds.), *Psychopharmacology in British Literature and Culture, 1780–1900*, Palgrave Studies in Literature, Science and Medicine,
https://doi.org/10.1007/978-3-030-53598-8_8

seeing did they inhibit? Every act of making something visible involves rendering other things invisible. De Quincey's text needs to be unpacked, not only to study the specific ways in which his historical moment understood opium but also to reveal ways of thinking about opium that have been rendered invisible in such canonical expressions of known drug effects within British literature. To do this, this paper will use a theoretical framework provided by Bruno Latour's Actor-Network-Theory (ANT).

The first section of this paper, titled "De Quincey, Opium and Anthropocentric Utilitarianism", will provide a close reading of *Confessions*, contextualising it through several medical texts on opium from the period in order to argue that *anthropocentric utilitarianism*, or the tendency to objectify entities in the world and evaluate them in relation to the advantage/disadvantages they have to human subjects, was the common trope through which nineteenth-century British medical and literary discourses understood opium. This is an orientation that has arguably become the dominant way of understanding opium in the modern Eurocentric world. The next section titled "Seeing Through the Eyes of Bruno Latour and Michael Pollan" will then establish a theoretical frame by synthesising Latour's ANT and Pollan's work at the intersection of botany and culture. Using this frame, then, the subsequent section "Opium and Plant-Human Co-Evolution" will read De Quinceyan representations against the grain and probe through the gaps and silences present within them, pointing towards possible new ways of understanding and representing opium that do not constrain the drug within the limited frame of a unilinear, utilitarian relationship between an active human subject and a passive opiatic object. Instead, I look at opium as an important variable with a mediating influence of its own that impacted the sociocultural networks of nineteenth-century Britain in significant ways. Finally, in "Looking Ahead", I will reflect on the possible future pathways that such new ways of researching the representation of drugs can open up.

De Quincey, Opium and Anthropocentric Utilitarianism

In order to better understand the rhetorical ways in which De Quincey framed opium, a good place to start is the narrative architecture of the text because it forms the conceptual frame through which readers enter De Quincey's world. While scholars have studied this narrative structure from many different angles (e.g. see J.H. Miller's analysis of *Confessions* as a text that demonstrates a crumbling biblical structure in a world bereft of God [1963], or see Platzner's evaluation of De Quincey's novel as a failure to adhere to the structural requirements of a confessional literary autobiography [1982]), surprisingly its entanglement in the utilitarian ethos of the time has not been given much attention, in spite of the very prominent presence of words like *pleasure* and *pain* in the section titles itself.

The narrative architecture of *Confessions of an English Opium-Eater* consists of two main sections which can be called the *Pre-Confession* section and the *Confession* section. Each of these sections, in turn, has two subsections. The *Pre-Confession* section consists of a *To the Reader* note, which is then followed by *Preliminary Confessions*. Both of these introduce readers to the persona of De Quincey in order to create "some previous interest of a personal sort in the confessing subject" (1821, 7). De Quincey does not shy away from claiming that "my life has been, on the whole, the life of a philosopher" (4) and then goes on to emphatically build up his persona as a "subtle thinker" who has "superb intellect" and "moral faculties" (7), thereby creating a solid anthropocentric fulcrum on which the rest of the narrative rests. This subject is then framed within the confines of a confessional mode—as someone who, in spite of having such impeccable intellectual and moral traits, has ended up becoming an opium-eater, thus "subject[ing] himself to such a yoke of misery; voluntarily to incur a captivity so servile, and knowingly to fetter himself with such a sevenfold chain" (De Quincey 1821, 6).

The rhetorical purpose of writing this text is then presented as the author's desire to absolve himself from the social ostricisation that he feels comes with not just eating opium but indulging in "it to an excess not yet recorded of any other man" (De Quincey 1821, 5), which he later quantifies as "fifty or sixty grains to 150 a day" (82). Instead of directly admitting guilt, however, De Quincey attributes a purposiveness to his text which he hopes will absolve him of any immoral acts that he may have committed: "Guilt, therefore, I do not acknowledge; and if I did, it is possible that I might still resolve on the present act of confession in consideration of the service which I may thereby render to the whole class of opium-eaters" (De Quincey 1821, 4). How can we understand this rhetorical move of writing a confession that is, at best, ambiguous about directly admitting guilt and instead seeks for absolution through the publishing of a text that will "prove not merely an interesting record, but in a considerable degree *useful* and *instructive*" (3 emphasis added)?

For Michel Foucault, the genre of confession is a "central component in the expanding technologies for the discipline and control of bodies, populations and society itself" (Dreyfus 1982, 174) as it involves an "interplay of confession, truth and power" (174) where the confessing subject's narrative of guilt is framed not in terms of an abstract or universal moral code but is entrenched very firmly within the epistemic and ethical paradigms of the historical moment that he or she belongs to. It is in line with this observation that we can begin to see utilitarian patterns of thinking within De Quincey's confession as he constantly stresses how "the *benefit* resulting to others from the record o f an *experience purchased at so heavy a price* might compensate, by a vast *overbalance*, for any violence done to the feelings I have noticed, and justify a breach of the general rule" (De Quincey 1821, 4 emphasis added).

These patterns become the most prominent in the second section of the text which is structured along a very explicitly utilitarian binary. This section consists of two subparts—the *Pleasures of Opium* and the *Pains of Opium*, which catalogue a phantasmagoric journey that the De Quinceyan persona underwent on coming in contact with opium. Oscillating between this De Quinceyan "subject" and the positive/negative effects that the "object" of opium had on it, this opium-eater's journey, which, in some ways, also contains elements of a *Bildungsroman*, charts the ways in which opium played a vital role in the growth of the autobiographical protagonist by affecting the aesthetic temperament of his mind as well as the biological constitution of his body.

While doing so, De Quincey constantly stresses the value of his text to several agents, the chief of which are opium-eaters: "[F]rom this account, rambling as it may be, it is evident that thus much of benefit may arise to the persons most interested in such a history of opium, viz., to opium-eaters in general, that it establishes, for their consolation and encouragement, the fact that opium may be renounced, and without greater sufferings than an ordinary resolution may support" (De Quincey 1821, 89). The text also suggests that the medical community stands to gain a lot from De Quincey: "[I]f the gentlemen of Surgeons' Hall think that any *benefit* can redound to their science from inspecting the appearances in the body of an opium-eater, let them speak but a word, and I will take care that mine shall be legally secured to them" (90 emphasis added). His overall assessment of opium, in line with this overarching logic of utilitarianism, remains limited to the idea that in spite of the several curiosities it inspires, the pains of opium outweigh its pleasures, which is why, at the end of this text, he stresses that "[t]he moral of the narrative is addressed to the opium-eater [...]. If he is taught to fear and tremble, enough has been effected" (De Quincey 1821, 82).

This utilitarian orientation not only frames De Quincey's assessment of his narrative but also tends to condition his understanding of the very effects of opium on his body, which he triangulates along an axis: with pain (negative) on one end and pleasure (positive) on another. On consuming opium for the first time, De Quincey writes: "That my pains had vanished was now a trifle in my eyes: this negative effect was swallowed up in the immensity of those positive effects which had opened before me – in the abyss of divine enjoyment thus suddenly revealed" (De Quincey 1821, 41). His assessment is further peppered with his understanding of the medical and commercial profitability of such a commodity:

> Here was a panacea, a φαρμακον for all human woes; here was the secret of happiness, about which philosophers had disputed for so many ages, at once discovered: happiness might now be bought for a penny, and carried in the waistcoat pocket; portable ecstacies might be had corked up in a pint bottle, and peace of mind could be sent down in gallons by the mail-coach. (De Quincey 1821, 41)

De Quincey's text on opium did not, of course, emerge out of a discursive vacuum. In spite of being quite unprecedented within the literary paradigm, it was very much informed by medical discourses that had been building up from over a century before his time and had become dominant in shaping the public's understanding of opium's medical effects when he was writing. De Quincey was, in fact, first introduced to laudanum (a mixture o f opium and alcohol) not for the purpose of inspiring any poetic sensibilities but simply as a medicine for his "excruciating rheumatic pains of the head and face" (De Quincey 1821, 40). If one looks at the popular medical texts of the times (some of which, in fact, De Quincey cites in his text), one starts to get a better understanding of his discursive genealogy, as these medical texts naturalised a system of thinking where opium was seen through the lens of anthropocentric utilitarianism.

At the beginning of his account, for example, De Quincey makes a reference to John Awsiter's *An Essay on the Effects of Opium*, which was written by an apothecary at the Greenwich hospital in 1810. At the beginning of this treatise, Awsiter presents his particular approach to the effects of opium by acknowledging that while there are "various shapes wherein it may be advantageously used for many diseases", he is more interested in "confin[ing] myself to that Property, which makes it considered as a poison" (Awsiter 1810, 4). He then goes on to elaborate on the painful effects of opium by making claims like opium brings "the animal Powers into immediate Conflict with the noxious Body, which produces Convulsions, and other dreadful Symptoms, the Forerunners of Death" (Awsiter 1810, vii). In this text, Awsiter makes constant references to Dr John Jones' *Mysteries of Opium Revealed*—an authoritative text on opium that, in spite of being published nearly a century earlier in 1701, had a huge impact on shaping writers like Awsiter, his predecessors as well as his peers. At various points in this almost hundred-year-long history of medicine between Jones and Awsiter, one sees the gradual emergence of an anthropocentric utilitarian gaze through which psychoactive substances were classified and evaluated. Akin to De Quincey's framing of opium's drug effects within utilitarian poles, medical practitioners like Awsiter, Jones and many others like them understood opium in relation to its "utility" for human bodies. For example, Jones too detested usage of opium in high doses, as he felt that it causes "[g]reat, and even intolerable Distresses, Anxieties and Depressions of Spirits, which in few days commonly end in a most miserable Death" (quoted in Davenport-Hines 2004, 45). If used in moderation, however, Jones felt that opium could be useful for things as diverse as gout, asthma, dysentery, vomiting, menstruation pains and even for increasing the size of breasts and penises (Davenport-Hines 2004, 44). In fact, he even goes on to recommend a mild dose of opium as he felt that it made people "more serene, and *apt for the Management of Business*, and *near dispatch of Affairs*" (quoted in Davenport-Hines 2004, 44 emphasis added), which, in fact, suggests the development of a proto-utilitarian vocabulary much before utilitarianism as a doctrine took hold in England.

Many other medical writers of the time helped popularise a similar vocabulary. John Brown was one such important figure. Even though De Quincey does not directly cite him, Brown's work was immensely popular with a range of Romantic thinkers like Coleridge, who in fact thought that he was of the same stature of men like Cicero, Milton, Shakespeare and Wordsworth (quoted in Vickers 1991, 48). Brown's main theory of Brunonianism, which found its most prominent articulation in his famous book *The Elements of Medicine* (1780), states that the human body is subject to a range of excitation from both internal and external stimuli, and that disorders are caused when there is either too much or too little excitation (Brown 1814, lxii–lxiii). Within this framework, Brown lauded opium by writing that "if anyone is sleepy without an evident cause, he will by opium be rendered surprisingly sprightly, lively, and vigilant; it banishes melancholy, begets confidence, converts fear into boldness, makes the silent eloquent, and dastards brave" (1814, 244). Simultaneously, he also warned that if used in excess and in combination with other substances, opium "may suddenly induce indirect debility, and even death by exhausting excitability" (244).

While Latin was still the dominant language of discourse within the medical community, writers like Awsiter had started to present their arguments "set down in English" without many "technical expressions" (Awsiter 1810, viii), or like Brown had been translated widely, as this was a time "when medical guides were being issued with commercial appeal to the reading public" (Davenport-Hines 2004, 43). Their enthusiasm was important in presenting opium as a lucrative commodity to the upcoming middle classes that could help them in experimenting and in fine-tuning their identities. Davenport-Hines writes that "in the eighteenth century there were innumerable quacks peddling nostrums which purported to transform human nature; they were the Age of Reason's equivalent of New Age gurus and herbalists of the late twentieth century, although they sometimes displayed the marketing ruthlessness of latter-day drug barons" (53). Multinational companies like East India Company, Bayer AG, Parke Davis and Co. and Winslow Curtis & Perkins, were, in fact, aided in their attempts to introduce opiatic products into the pharmaceutical market by texts like those of Jones, Awsiter, Brown and De Quincey, as these texts, whether intentionally, or unintentionally, acted as advertisements for them.

Seeing Through the Eyes of Bruno Latour and Michael Pollan

If we now look at this intertextual tapestry of texts on opium, the larger epistemic categories and discursive tropes through which opium use was understood across literary and medical genres in eighteenth- and nineteenth-century Britain start to emerge. At a very elemental level, all these texts straightjacket the interaction between humans and opium by first framing it as a linear action between an agential human subject and an inert object. Such texts then proceed to study this restricted, linear action in various ways, with the overall purpose being to determine the use-value of this action (i.e. to determine whether opium, as a commodity, was beneficial or not *vis-à-vis* its effects on the human body).

The context for this action always remains anthropocentric, and all other actors and material contexts that may be involved in this equation, like the botanical and material origins of opium and its chemical constitution, or the colonial systems of production and distribution through which they emerged, are either only briefly referred to, or are completely brushed under the carpet. They are never thought of as important variables that need to be taken into consideration while assessing the overall utility of opium consumption. Mired within the logic of anthropocentrism and utilitarianism, these medical texts have formalised a certain commoditised way of understanding drug consumption, or formed what Kenneth Burke would call a "terministic screen", a screen composed of nomenclature that "directs the attention into some channels rather than others" (Burke 1966, 45). In its anthropocentric utilitarian orientation, the De Quinceyean terministic screen has rendered invisible a whole spectrum of actors that are intimately involved in all acts of drug consumption.

At the heart of this terministic screen lies a metaphysical assumption—a circumscribed understanding of the term *action* as that which only flows causally and in an unidirectional manner between a subject and an object, where the subject is always human, and the action is something that should simply be assessed in terms of the overall utility that it produces. Bruno Latour in his *Reassembling the Social* critiques such epistemic castrations of action and provides us with a new paradigm to rethink action at a very elemental level. Instead of being thought of as limited between a subject and an object, Latour argues that "action should rather be felt as a node, a knot, and a conglomerate of many surprising sets of agencies that have to be slowly disentangled" (Latour 2009, 44). To achieve this disentanglement of action, Latour suggests the use of a theatrical metaphor and prescribes the framing of any action within an expansive actor-network framework:

> An 'actor' in the hyphenated expression actor-network is not the source of an action but the moving target of a vast array of entities swarming toward it. [...] To use the word 'actor' means that it's never clear who and what is acting when we act since an actor on stage is never alone in acting. Play-acting puts us immediately into a thick imbroglio where

> the question of who is carrying out the action has become unfathomable. Does the audience's reaction count? What about the lighting? What is the backstage crew doing? Is the playwright's message faithfully transported or hopelessly bungled? Is the character carried over? And if so, by what? What are the partners doing? Where is the prompter? If we accept to unfold the metaphor, the very word actor directs our attention to a complete dislocation of the action, warning us that it is not a coherent, controlled, well-rounded, and clean-edged affair. By definition, action is dislocated. Action is borrowed, distributed, suggested, influenced, dominated, betrayed, translated. (Latour 2009, 46)

By following Latour's recommendation, we begin to see that in the act of De Quincey consuming opium, it is not simply a linear, causal relationship between De Quincey and the piece of opium that can be assessed in terms of a simplistic understanding of utility. What exists instead is a complex, multi-layered actor-network populated by a wide range of human and non-human actors that needs to be studied beyond a simplistic framework of utility. We must keep in mind that opium is not an abstract, inert substance that comes out of nowhere. It has a dynamic, botanical-chemical constitution that has emerged due to thousands of years of coevolution between plants and humans, and the substance has been mediated by complex sociocultural and political forces.

All acts of drug consumption, in fact, clearly lie at the intersection of the botanical, the biochemical and the sociocultural domains. A major methodological problem, however, that prevents us from studying such complexly entangled phenomenon lies in the modern academic system, which has constructed botany, chemistry and culture as separate disciplines, to be studied by people separately trained in them. This is why there are very few methodological tools by which one can engage with these various dimensions together. However, new interdisciplinary ways of studying drugs have begun to emerge.

Michael Pollan's pioneering work at the intersection of botany and culture helps one to contextualise opium usage within a larger planetary history of plant-human coevolution and ponder over ways in which we can re-read De Quincey's text within such a planetary history. Through this act of decentring De Quinceyan aesthetics of intoxication along an actor-network framework, we can rethink epistemic categories that have been intrinsically tied to British psychopharmacological discourse, categories which have arguably played a major role in defining the global, modern discourse on drugs given Britain's colonial as well as politico-economic control over much of the globe in the nineteenth as well as the early twentieth centuries. The limits of this paper dictate that these explorations can only be initial brushstrokes, but they are important nonetheless, as they will hopefully chart out new paths for future studies.

Opium and Plant-Human Coevolution

De Quincey begins the *Pleasures of Opium* section of his account by fetishising his first act of opium consumption thus: "I feel a mystic importance attached to the minutest circumstances connected with the place and the time and the man (if man he was) that first laid open to me the Paradise of Opium-eaters" (1821, 40). This man, De Quincey goes on to write, was a "druggist – unconscious minister of celestial pleasures [who] has ever since existed in my mind as the beatific vision of an immortal druggist, sent down to earth on a special mission to myself" (40). In this description, opium seems to have come in a supernatural manner straight from the "immortal druggist" himself. The narrative imagination claims to attach importance to the "minutest circumstances" related to the first act of opium consumption, but these circumstances remain limited to the druggist who is but one link in the complex chain of opium production that begins from the natural realm (i.e. with the *Papaver somniferum* plant). Similarly, most medical texts of the period did not engage with the botanical origins of opium much, apart from giving it a cursory glance to give their readers an abrupt, mechanistic run-through of the process involved in extracting opium from the poppy plant. Why were opium's botanical origins not paid much attention to by eighteenth- and nineteenth-century British literary and medical texts?

Wandersee and Schussler refer to such ignorance of the botanical variables involved in shaping human culture as "plant blindness", a condition integral to modern anthropocentric terministic screens which create an "inability to see or notice the plants in one's own environment, leading to: (a) the inability to recognize the importance of plants in the biosphere, and in human affairs; (b) the inability to appreciate the aesthetic and unique biological features of the life forms belonging to the Plant Kingdom" (Wandersee and Schussler 2001, 3). If one were willing to move beyond one's "plant-blindness", one would start to realise that the "plantness" of opium is pivotal to understanding human-opium interactions, something that many eighteenth- and nineteenth-century British writers across literary and medical domains missed.

The rigid anthropocentric separation between nature and culture that one notices in De Quincey's, Awsiter's, Brown's and Jones' unawareness of the "plant-ness" of opium can be better understood by using recent advancements in the fields of anthropology, ecology and ethno-epistemology b y scholars like Phillip Descola, who have historicised this separation as a unique trait of Anglo-European ways of thinking and being rather than as a fundamental characteristic of the world itself. Several non-Anglo-European communities, like the Achuar tribe of the Amazon whom Descola studies, have a completely different symbolic order which frames the world around them not in terms of a sharp division between nature and culture, but in terms of a "great social continuum that includes both humans and non-humans" (Descola 2014, 7), which allocates an important sense of humanlike "being-ness" even to the non-human actors in their network.

Pollan, a twenty-first-century botanist, applies similar ways of seeing to create what he calls a "plant's eye view of the world" (2002, 21), in which he attempts to move beyond modern plant blindness to look at the botanical variables intimately involved in most human-drug interactions. In an essay called "Opium Made Easy" published in 1997, Pollan questions the highly unlikely probability of a molecule produced by the poppy plant to randomly hold the precise key that helps unlock the physiology of pleasure and pain in human brains. He argues that while it could be an accident, it is more likely that it is a result of coevolution:

> [O]ne theory holds that Papaver somniferum is a flower whose evolution has been directly influenced by the pleasure, and relief from pain, it happened to give a certain primate with a gift for horticulture and experiment. The flowers that gave people the most pleasure were the ones that produced the most offspring. It's not all that different from the case of the Bourbon rose or the beefsteak tomato, two other plants whose evolution has been guided by the hand of human interest. (1997, n.pag.)

Common sense, however, continues to dictate that we see such processes of coevolution only in terms of anthropocentrism because the very grammar of our languages forces people to imagine the relationships between humans and plants in terms of master–slave metaphors where "I choose the plants, I pull the weeds, I harvest the crops" (Pollan 2002, 8). But what if this were, at least in part, a cultural delusion?

Moving beyond master–slave metaphors, Pollan suggests the use of a different metaphor, one that helps us visualise plant-human interactions in terms of a non-hierarchical, coevolutionary relationships like the one symbolised by a flower and a bee, where both parties act to pursue their respective interests but cumulatively end up engaging a cooperative form of trading where the bee gets food and the plant's genes get transportation (Pollan 2002, 8–9). Just like the bee and the flower, the relationship between plants and humans, too, is symbiotic rather than unidirectional. Coming specifically to the biochemical nature of this relationship, Pollan writes that plants' evolution, unlike that of most animals, has focused on making them excel in the art of biochemistry, of producing and using chemicals in order to interact with other creatures in their environment in a wide range of ways, including defence, offense, nutrition and intoxication:

> Some plant toxins, such as nicotine, paralyze or convulse the muscles of pests who ingest them. Others, such as caffeine, unhinge an insect's nervous system and kill its appetite. Toxins in datura (and henbane and a great many other hallucinogens) drive a plant's predators mad, stuffing their brains with visions distracting or horrible enough to take the creatures' mind off lunch. Compounds called flavonoids change the taste of plant flesh on the tongues of certain animals, rendering the sweetest fruit sour or the sourest flesh sweet, depending on the plant's designs. (2002, 76)

A key part of these coevolutionary processes between plants and humans has been the development of cultures of consumption around different psychotropic substances at different points of time. These processes are not as arbitrary as they might seem at first, but come into being through an interaction between a wide range of material factors that enable the selection of a drug within a culture, as well as condition the interpretation of its effects.

If we were to contextualise De Quincey's novel within this larger history, an interesting new set of observations start to emerge. Opium became prevalent in the eighteenth and nineteenth centuries not because of a random twist of fate, but because certain material, political and sociocultural forces created an environment that enabled the evolutionary selection of opium in Britain to an unprecedented degree. It is in this historical phenomenon that we can see how the botanical and chemical constitution of the poppy plant mediated cultural expressions within the nineteenth-century public sphere.

One possible reason for opium's selection in this period was perhaps the perception of some commentators of its depressant and analgesic qualities that helped to counter the cognitive effects caused by an industrial culture where speed was fast becoming the dominant behavioural motif. While some stimulants helped people integrate themselves into the painful drudgeries of fast-paced industrial cities, the pleasures o f opium provided an alternate universe of calm, imagination and repose, which is why opium became so central to a range of Romantic writers and poets whose aesthetic reveries often revolved around its consumption. This contrast mediates the effects of opium in De Quincey's account very prominently, as he writes that in consuming opium, "it seemed to me as if then first I stood at a distance and aloof from the uproar of life; as if the tumult, the fever, and the strife were suspended; a respite granted from the secret burthens of the heart; a sabbath of repose; a resting from human labours" (De Quincey 1821, 51). Furthermore, he writes that in highest states of opiatic ecstasy, "markets and theatres are not the appropriate haunts of the opium-eater" as "crowds become an oppression to him" which is why he "naturally seeks solitude and silence, as indispensable conditions of those trances, or profoundest reveries" (De Quincey 1821, 50).

Apart from providing immense neural pleasure, De Quincey also claimed that opium had hallucinogenic effects which helped counter the mundane material realities of the growing urban and industrial ethos of the times:

> [T]hou buildest upon the bosom of darkness, out of the fantastic imagery of the brain, cities and temples beyond the art of Phidias and Praxiteles—[...] and from the anarchy of dreaming sleep callest into sunny light the faces of long-buried beauties and the blessed household countenances cleansed from the dishonours of the grave. Thou only givest these gifts to man; and thou hast the keys of Paradise, oh, just, subtle, and mighty opium! (De Quincey 1821, 52)

The two poles implied in this equation—erratic industrialisation and opiatic reveries—are not entirely antithetical to each other, as they might seem at first. They are, in fact, two sides of the same coin, tied together by a language of profit and utilitarian capitalism. Since opium helped numb some of the drastic effects of urbanisation and industrialisation, its pleasures were framed in texts like De Quincey's as an antidote for the side effects of capitalism. This is in fact a connection that exists even today. Opium in its raw tincture form might not be as accessible as it was in De Quincey's time, but most pharmaceutical companies still use the substance as the primary agent in a range of psychopharmacological medicines for pain management, and in some cases even for depression and anxiety (Tenore 2008), as well as for other mental health issues that have been found to have a strong correlation with practices engendered by capitalism (Tweedy 2017). The current opium crisis seen in many parts of the world in the USA and India is indeed a fallout of this complex entanglement that opium has had with Western capitalism (Muhuri et al. 2013; Gupta 2016; Pereira 2018).

Looking Ahead

The purpose of this paper has been to re-read representations of opium in nineteenth-century British literature by opening them up to the criss-crossing impulses of ANT. Such a reading allows us to think of opium not as a passive object caught up in a utilitarian relationship with human subjects, but as an active botanical-chemical actor that mediated the socio-cultural networks of nineteenth-century Britain and beyond—an episode that was a pivotal historical moment within an overarching coevolution in which the human species has been intimately tied to the poppy plant for thousands of years. The prevalent ways of understanding our relationship with opium (and drugs in general), however, continue to be coloured by anthropocentric and utilitarian frames of reference. Opium as a plant has thrived across the modern world primarily because it contains alkaloids that produce effects that have the potential to counterbalance certain behavioural and psychological effects, many of which are produced by capitalism. Blinkered by the paradigm forged at the time that De Quincey and his peers were writing about this substance, we have today overexploited opium, which has, in part, led to the addiction and trafficking crises we now see.

Through such a critique of elemental epistemic categories present in eighteenth- and nineteenth-century British literary and medical discourses, what is being proposed here is a call for a new way to understand human-drug interactions—a paradigm of writing that cuts across disciplines and re-imagines these interactions in line with an actor-network framework. At the heart of ANT lies an attempt to create a sort of empathy between the several actors that populate a social system (Latour 2009, 259–62). This paper has been written with such a hope of initiating a new historically oriented understanding of drugs that does not only prioritise the user but is also conscious and respectful of the "beingness" of the botanical and cultural variables intimately involved in all acts of drug consumption.

References

Awsiter, John. 1810. *An Essay on the Effects of Opium: Considered as a Poison; with the Most Rational Method of Cure, Deduced from Experience; Directing Likewise the Proper Means to Be Used When Physical Assistance Cannot Be Readily Obtained; Necessary to Be Universally Known, for the Preservation of Life*. London: G. Kearsly.

Brown, John. 1814. *The Elements of Medicine: Or, a Translation of the Elementa Medicinae Brunonis*, ed. Thomas Beddoes. Philadelphia: James Webster.

Burke, Kenneth. 1966. *Language as Symbolic Action*. Berkeley, Los Angeles and London: University of California Press.

Davenport-Hines, Richard. 2004. *The Pursuit of Oblivion: A Global History of Narcotics*. New York: W. W. Norton & Company.

De Quincey, Thomas. 1998 [1821]. *Confessions of an English Opium-Eater*. Oxford: Oxford University Press.

Descola, Philippe. 2014. Configurations of Continuity. In *Beyond Nature and Culture*, trans. Janet Lloyd, 3–31. Chicago: University of Chicago Press.

Dreyfus, Hubert L. 1982. *Michel Foucault: Beyond Structuralism and Hermeneutics*. Chicago: University of Chicago Press.

Gupta, Anuj. 2016. Punjab Is Flying High, But the Wings Came from the British Raj. Scroll.in. scroll.in/magazine/811102/punjab-is-flying-high-but-the-wings-came-from-the-british-raj. Accessed July 21 2019.

Jones, John. 1701. *The Mysteries of Opium Revealed*. London: Richard Smith.

Latour, Bruno. 2009. *Reassembling the Social: An Introduction to Actor-Network-Theory*. Oxford: Oxford University Press.

Miller, Joseph Hillis. 1963. *The Disappearance of God: Five Nineteenth Century Writers*. Oxford: Oxford University Press.

Muhuri, Pradip, et al. 2013. Opioids and Capitalism. CBHSQ Data Review. www.samhsa.gov/data/sites/default/files/DR006/DR006/nonmedical-pain-reliever-use-2013.htm. Accessed February 20 2020.

Pereira, Paulo. 2018. Opioids and Capitalism. Centre for International Policy Studies. www.cips-cepi.ca/2018/07/10/opioids-and-capitalism/. Accessed February 1 2020.

Platzner, Robert. 1982. De Quincey and the Dilemma of Romantic Autobiography. *Dalhousie Review* 61 (4): 605–17.

Pollan, Michael. 1997. Opium Made Easy. *Harper's Magazine. Harper and Brother's*, April 1.

———. 2002. *The Botany of Desire: A Plant's Eye View of the World*. New York: Random House.

Tenore, Peter L. 2008. Psychotherapeutic Benefits of Opioid Agonist Therapy. *Journal of Addictive Diseases* 27 (3): 49–65.

Tweedy, Rod. 2017. A Mad World: Capitalism and the Rise of Mental Illness. *Redpepper*, August 9.

Vickers, Neil. 1991. Coleridge, Thomas Beddoes and Brunonian Medicine. *European Romantic Review* 8 (1): 47–94.

Wandersee, James H., and Elisabeth E. Schussler. 2001. Towards a Theory of Plant Blindness. *Plant Science: The Botanical Society of America Spring* 47 (1): 2–8.

CHAPTER 9

Blood Streams, Cash Flows and Circulations of Desire: Psychopharmacological Knowledge About Opium in Nineteenth-Century Women's Fiction

Nadine Böhm-Schnitker

For Wilkie Collins, the completion of his sensation-*cum*-detective novel *The Moonstone* required a "physical sacrifice" (2008, lv). "[C]rippled in every limb by the torture of rheumatic gout" and stricken by his mother's death, "Collins imbibed huge amounts of laudanum" (Sutherland

[1] Alcohol and opium, though entirely different in their effects on the body (see Peters 1981, 469–70), were frequently considered together, both being stimulants in the Brunonian sense; in addition, they were associated through the concept of inebriety that had an impact on the development of "the British version of the disease model" of addiction, as it had "close links with temperance and anti-opium ideology" (Berridge 1978, 457).

N. Böhm-Schnitker (✉)
English and American Studies, Friedrich-Alexander-Universität Erlangen-Nürnberg, Erlangen, Germany
e-mail: nadine.boehm-schnitker@fau.de

N. Roxburgh and J. S. Henke (eds.), *Psychopharmacology in British Literature and Culture, 1780–1900*, Palgrave Studies in Literature, Science and Medicine,
https://doi.org/10.1007/978-3-030-53598-8_9

2008, xxxv)—a dissolution of opium in alcohol[1]—in order to be able to continue working. Mindful that he "had his duty to the public still to bear in mind" (lv), he overcame both his physical and his mental suffering, so that the next instalment of *The Moonstone* was ready for public consumption and serial publication in Charles Dickens' journal *All the Year Round*. The palliative function of opium and the smooth functioning of the economy are intertwined in Collins' endeavour. Correspondingly, in the contemporary literary criticism, the production and consumption cycles of sensation fiction have been likened to a form of addiction early on.[2] The conservative critic Henry Mansel represents a key example. In his opinion, sensation fiction plays a pivotal role in

> moulding the minds and forming the habits and tastes of its generation; and doing so principally, we had almost said exclusively, by 'preaching to the nerves'. [...It is] called into existence to *supply the cravings of a diseased appetite*, and contributing themselves to foster the disease, and to *stimulate the want which they supply*. [...] No divine influence can be imagined as presiding over the birth of [the sensation writer's] work, beyond the *market-law of demand and supply*. (Mansel 1863, 495–96, emphasis added)

Mansel creates an important discursive connection between market laws and the consumption of 'addictive' literature that satisfies quasi-physical cravings and thus brings together capitalism, culture and drugs and employs a particular strand of medical discourse to articulate the different aspects. He draws on a Brunonian theory of stimulants—sensation novels 'stimulate' the demand for ever more sensations, he argues—and within this genre, literary stimulation has clearly gone rampant, unbalancing a healthy equilibrium.

John Brown's medical theory "permeates popular medical thinking in the first half of the nineteenth century to a surprising extent, perhaps because Brown's radical simplification of the causes of disease into over-stimulation and understimulation could easily be assimilated to ideas about the nervous origins of disease originating with Brown's teacher William Cullen" (Budge 2012, 12). The sensational thrills of sensation fiction can easily be grouped among stimulants and, with habituation, this theory already gestures towards notions of addiction since "different

[2] See essay in this volume by Sarah Frühwirth.

kinds of stimuli – whether of mind or body – could have similar effects when experienced through a 'habit' of being 'over stimulated'" (Colman 2019, 6). While addiction as such is a notion that emerges only in the 1870s, symptomatically after the 1868 Pharmacy Act that restricted the sale of opium to pharmacists (see Peters 1981, 455–56), Collins' *The Moonstone* tackles some issues of addiction such as drug dissociation and drug tolerance *avant la lettre* and supposedly represents "the first psychopharmacological thriller" (Siegel 1983, 161–62; 167; 175).

As Susan Zieger has shown, drug consumption also symptomatically uncovers colonial power relations since "[h]abit-forming substances reveal the imperial dependence on colonial resources and labour – a dependency felt physiologically, at the intimate physical level of craving – and yet forgotten most easily in the effects of oblivion" (2011, 211). This is a discursive and material triangle marked by several dependencies: a dependence on substances such as opium, on labour and economic contexts and, with them, on colonial expansion. This triangle proves both quite enduring (see Angell 2018, 56; Haiven 2018, 663) and is not without precedent. Taking my cue from Collins' allegedly first psychopharmacological novel, I wish to explore the psychopharmacological knowledge encoded in early nineteenth-century novels by women writers. I will focus on an 'addiction novel' (see McAllister 2015, 301), Maria Edgeworth's *Belinda* (1801), an industrial novel, Elizabeth Gaskell's *Mary Barton* (1848), and will provide an outlook on Harriet Martineau's *Illustrations of Political Economy* (1832–34), in which she discusses opium both as a drug and as an economic commodity in a transnational context. The two novels feature characters that depend on opium for very different reasons, but both of them reflect on the known drug effects in a social, economic and colonial context, which turns them into crucial literary precursors for later psychopharmacological fictions.

Brunonianism provides an apt medical framework for the ways in which these novels deal with opium consumption in particular; even advanced pharmacological knowledge about opium in the later nineteenth century continues to be understood within this particular framework of medical knowledge. Dolores Peters pertinently highlights that "improved pharmacological knowledge – of the chemical action of alkaloids, for example – seemed more likely to become entangled in Brunonian terminology than to lead to an understanding of addiction that was useful to practitioners" (1981, 456). With Gavin Budge, one may consider Brunonianism "a cultural meme" (2012, 14) that remains productive well into and even beyond the mid-Victorian age despite its many vilifications "as mere quackery" (Bynum and Porter 1988, ix). In the introduction to the first volume of his *The Elements of Medicine*, Brown defines life as "*a forced state*; if the exciting powers are withdrawn, death ensues as certainly as when the excitability is gone", while such exciting powers are classified as either external, such as "heat, food, wine, poisons, contagions, the blood, secreted fluids, and air", and internal, such as "the functions of the body itself, muscular exertion, thinking, emotion and passion" (1795, I cxxvii). Among external stimulants, Brown considered opium in any of its forms "sufficient for most purposes of high stimulation" (1795, II 15),[3] and as such it features prominently in Edgeworth's *Belinda*, while Gaskell's *Mary Barton* focuses on its employment against hunger. According to John Brown, health and illness are a function of the degree of excitement:

> Excitement, the effect of the exciting powers, the true cause of life, is, within certain boundaries, proportional to the degree of stimulus. The degree of stimulus, when moderate, produces health; in a higher degree it gives occasion to diseases of excessive stimulus; in a lower degree, or excessively weak, it induces those that depend upon a deficiency of stimulus, or debility. And, as excitement is the cause both of diseases and perfect health; so that which restores the morbid to the healthy state, is a diminution of excitement in diseases of excessive stimulus, and an increase of excitement in diseases of debility. These intentions are called Indications of Cure. (Brown 1795, I 14–15)

Correspondingly, health is considered to be a question of the 'management' of stimuli, medical intervention "the proper management of the stimulant plan of cure" (I 221). Such management of the body neatly ties in with the management of the mind required by pastoral forms of power and the risk management required by biopower in a Foucauldian sense (see Foucault 1982, 213–15). Ultimately, Brunonianism offers an "economic model of the body" in that it "emphasizes the limited supply of vital energy for which the brain is in competition with other bodily organs" (Budge 2012, 183). It "regarded all disease as a manifestation either of the intrinsically addictive condition of bodily over-stimulation, or of an understimulated condition resulting in the blockage of bodily

[3] The stimulating effect of opium was hotly debated, as the drug could also have sedative and narcotic effects (see Berridge 1999, xxi; 64; 66).

vitality" (Budge 2012, 25). It is thus an ideal flow that is to be upheld, and health depends on the amount and quality of stimulants that circulate in a body's blood stream.

The novels under consideration here not only focus on the gendering of different forms of opium consumption, but they render it abundantly clear that opium correlates with economic as well as colonial aspects that consequently interconnect the management of the body with the management of finances and colonial expansion. The Orientalism inscribed in founding examples of Romanticist opium literature such as Samuel Taylor Coleridge's "Kubla Khan" or Thomas De Quincey's *Confessions of an English Opium-Eater* has been widely explored and already indicates the literary negotiation of economic, colonial and physiological dependencies. In contrast to these two founding texts, the interrelations of women's writing and opium are less well explored. Taking my cue from Hannah Cowles Freeman's analysis of the way women writers focus on domestic uses of opium rather than on orientalised contexts like their male counterparts[4] (see 2012, 15), I will have a closer look at early nineteenth-century writers with a special focus on the discourse on opium's psychopharmacology and its politics at the dawn of the First Opium War. In this context, the domestic and the colonial are always already intertwined so that women's writing about the domestic use of opium cannot be read without perceiving its transnational reverberations. Opium circulates in blood streams as well as in economic channels, and the logic of these circulations intersects with the social construction of gender, race and class.

Nineteenth-century women writers document these intersections and betray the sociopolitical workings of opium by showing what kinds of psychological and physical relief the drug grants. Edgeworth's *Belinda*, for instance, foreshadows the concerns of later sensation fiction writers and documents the ways in which the psychological impact of opium betrays the intertwinements of the economic desires of empire, the cravings of the body, the structures of gendered and racial suppression and political economy. Belinda's somewhat problematic mentor figure, Lady Delacour, knows how to employ opium in order to play a socially acceptable role in public, while her private, domestic life is riven by her attempt to hide physical and mental pain through her secret consumption of

[4] See essay in this volume by Joseph Crawford.

laudanum. The deliberate usage of the drug thus becomes part of a cultural articulation of femininity. Gaskell's *Mary Barton* connects the political with the private and reveals opium addiction to be a crucial relay between individual desire, empire and domesticity. During the 1840s, it is hunger in particular that is battled with opium, and the hunger induced by the textile industry's economic downturn during the Opium Wars is a potent physiological marker of the intertwinements of the colonial and the domestic. Finally, Martineau spells out such discursive intertwinements more explicitly in her *Illustrations of Political Economy*, a popularisation of political economy that reflects on opium's importance for Britain's domestic economy. These texts provide a double reflection on psychopharmacology in that they represent the characters' knowledge about opium's effects on mind and body on the diegetic level and reveal the wider sociocultural contexts in which the drug plays a role.

Maria Edgeworth's *Belinda* (1801): The Spectacle of Femininity as Opioid Façade

Maria Edgeworth "devotes more o fher novel to representing addiction than any novel preceding this one" (Teynor Donatini 2011, 259; fn. 25). Her novel *Belinda* (1801) describes its eponymous heroine's entry into society under the tutelage of the notorious Lady Delacour, whose social behaviour is reformed by Belinda, in turn. After some errors—Belinda becomes engaged to the wrong man[5] before she marries 'Mr Right',

[5] The 'wrong' man Belinda is engaged to is Mr Vincent, "a creole" who made his fortune in Jamaica. He brings along "a black servant of the name of Juba", who is taught to stop believing in obeah, a belief that also comprises "medical practice" (Murison 2015, 144). While Juba marries an English woman, the marriage between Mr Vincent and Belinda is forestalled due to his gambling habit. Belinda then finally settles on the English Clarence Hervey. In a Brunonian context, it is decidedly Mr Vincent's West Indian origins that curtail his development due to a lack of stimuli, an ideology, for example, derived from Benjamin Mosley's A *Treatise on Tropical Diseases* (1787), Bryan Edwards' *The History, Civil and Commercial, of the British Colonies in the West Indies* (1793–1801) or Richard Reece's *Medical Guide, for Tropical Climates* (1814): "Mr. Vincent's promiscuous interactions with enslaved Africans and other creole children led to his appetite for gambling. [...] Mr Vincent's gaming and his inability to halt this vice are both ascribed to his West Indian upbringing and, by logical extension, to the climate of the tropics" (151). His West Indian upbringing hence leads to a lasting change that prevents him from achieving health in the sense of moderate stimulation—he cannot shake the habit of gambling—and thus his difference from other suitors becomes racialised within

and Lady Delacour needs to shake her desire for admiration that goes hand in glove with her dependence on opium, and gets her to understand that happiness lies within the confines of family life rather than outside of it. *Belinda* can be said to be *doing* psychopharmacology, as the whole novel cannot only be understood as a form of medical treatment but also presents novel consumption as such as a kind of drug that performs a mood change in its readers; *Belinda* performs medical advice in the process of being read, and the piece of advice it strives to transmit is moderation (see McAllister 2015, 321), particularly with regard to stimulus management; thus, the novel clearly subscribes to a Brunonian notion of health. It relies on a "particular storytelling situation that resembles the process of medical diagnosis during the time *Belinda* was written" (Teynor Donatini 2011, 248). Belinda's role as Lady Delacour's protegé is to serve as "a doctor diagnosing and treating a patient" and it is important to keep in mind that "this relationship historically was predicated on reciprocity and conversation, not just one-way examination" (248). Since Lady Delacour's "body is the source of narration" (248), her psycho-physiological dependence on opium becomes the main productive force for the narrative and thus *Belinda* represents a central example of "addiction literature" (McAllister 2015, 301). As already stated, the notion of addiction does not develop until the 1870s, hence the novel negotiates emerging views on habituation—including "habits of all kinds, whether they involve drug abuse, gambling, love, or novel reading" (Teynor Donatini 2011, 265). Lady Delacour develops from a flamboyant and seemingly selfish woman of society to a responsible mother and wife content in her domestic surroundings, and the reader is included in her learning curve in the process of reading through the quasi-medical conversation with her and the other characters. Indeed, the novel establishes a close connection between the internal and external communication systems of the novel—fictional characters and real readers need to learn the same lesson: to shake the habit of overstimulation, whether the latter is caused by the consumption of drugs, novels or other stimulants. What this pedagogical structure achieves is a normalisation and

the context of tropical medicine and renders him an unsuitable partner for Belinda in a Brunonian context. This can be extended to the money derived from Mr Vincent's Jamaican sugar plantation, which cannot enrich Belinda without associating her with Mr Vincent's incompatible health condition.

naturalisation of patriarchal structures and the heteronormative nuclear family.

The logic behind Lady Delacour's medical treatment and the path to what is shown to be healing conforms to what Michel Foucault has called pastoral power. This is a power form that "the modern Western state has integrated in a new political shape" and that "originated in Christian institutions" (Foucault 1982, 213). Its aim is "to assure individual salvation in the next world" by looking after "each individual in particular"; this aim "cannot be exercised without knowing the inside of people's minds, without exploring their souls, without making them reveal their innermost secrets" (214). According to Foucault's theory, this form of power came to be secularised around the eighteenth century and was then transformed from a care for salvation into health care and policing of the body. Lady Delacour comes into view as an individual woman whose self-fashioning is revealed to be an *egoistical* error (see Teynor Donatini 2011, 265) that can only be healed by way of confession and devoting herself to her family. Community, interrelation and openness are healing forces against individual egoism and habitualised dependencies. Tying in with Brunonian theory, the novel thus presents egoism as a blockage of vital energies, while the active flow of exchange restores health. It is Belinda who serves as a confessor and as a healing force for Lady Delacour, and with confession as the central practice of pastoral power, Belinda is able to dissolve the blockage the Lady suffers from.

Once admitted to Lady Delacour's well-guarded boudoir, "Belinda, as she looked round, saw nothing but a confusion of linen rags – vials, some empty, some full – and she perceived that there was a strong smell of medicines" (Edgeworth 2008, 31). The very atmosphere of Lady Delacour's private chamber is replete with drugs and drug paraphernalia. Apart from the maid Marriott, Belinda is the only character who is granted access to this private room, where she is witness to several revelations. The first is Lady Delacour's "deathlike countenance" that is normally covered by a thick layer of make-up (Edgeworth 2008, 31). The second is the cause of that appearance, an ailment of her breast that ensued from a gun backfiring in a female duel. Symptomatically, Lady Delacour's breast is hurt during a practice that is coded masculine and the whole enterprise is described as an attack against her femininity. Her daring to partake in such a masculine undertaking is punished by an injury to the feminine breast that can only heal once she admits an actual doctor to examine her, a process which is accompanied with re-establishing the family ties to her husband and her daughter. "In a novel full of parallels between physical and emotional or social ailments, the supposedly clogged and festering glands in Lady Delacour's breast offer an apt figure of her egoism, in which the psyche blocks a free-flowing exchange of ideas and emotions with others" (Teynor Donatini 2011, 266). Such an equation of health with both the free flow of ideas and emotions as well as with the social exchange within the nuclear family once more illustrates the novel's reliance on Brunonian theories on health that similarly depend on a physiological equilibrium. They "identify vital energy both with the circulation of the blood and with nervous fluid" (Budge 2012, 2). Lady Delacour's performance of femininity that she can only maintain when separated both from her husband and her daughter is thus clearly identified with an unhealthy blockage that must be held in check by stimulants such as opium; without them, this blockage would lead to her death. Free, uninhibited blood streams are markers of health in *Belinda*, as is financial liquidity. While Lady Delacour's flamboyant lifestyle leads her into debt, her financial situation is balanced once she returns to her family life.

In the end, Lady Delacour can be healed by reforming her values and re-entering the community (see McAllister 2015, 323), a view which mirrors contemporary medical knowledge: "That Lady Delacour's breast cancer appeared and disappeared so easily indicates that Edgeworth was aware of the accepted medical concept of the time that breast cancer could be both caused and cured by changes in the patient's emotional state" (Larson 2009, 198). Opium, it seems, can only bring about an artificial and temporary change in Lady Delacour's emotional state, but her return to stereotypes of femininity—associated with domesticity, community and the family—can effect a lasting change to the better. Even though this plot turn is reflected upon and metafictionally exhibited along with the characters in the final tableau (see Edgeworth 2008, 478), these healing forces remain unchallenged and thus underwrite the novel's conservative gender politics.

Another dependency Lady Delacour forms to battle her condition is to resort to religious discourses. However, the heterodiegetic narrator is quick to point out that Lady Delacour's newfound faith in Methodism is rather a withdrawal symptom than her actual conviction. What is more, Methodism can indeed be understood as a possible substitute for opium in a Brunonian context, since "Methodist belief can only be kept up by a perpetual nervous overstimulation which accounts for the hysterical character of the Methodist conversion experience" (Budge 2012, 25). Just like opium, Methodism is a central stimulant that Lady Delacour requires to maintain a certain state of health.

Popular knowledge in the early nineteenth century comprised the view that opium could be employed "as a remedy for the 'fatigue and depression' unavoidable in working-class life at the time" (Berridge 1999, 31), but equally relevant for Lady Delacour's lifestyle.[6] It is decidedly *performance* that defines Lady Delacour, the theatre and opium. As long as they all do perform, the show can go on. Once they cease to perform, the ghosts held in check through performing reappear and 'obtrude' on the peace granted by the drug or Methodism. Such stimulants are required for the smooth running of the Lady's social charade with all the collateral damage entailed by it, and, of course, this smooth running also requires economic liquidity to finance it (see Edgeworth 2008, 39). Secret terrors resurface to haunt Lady Delacour as 'side-effects' of her lifestyle, which she herself describes as a kind of show to Belinda:

> Ambitious of pleasing universally, I became the worst of slaves – a slave to the world. Not a moment of my time was at my own disposal – not one of my actions; I may say, not one of my thoughts, was my own. I was obliged to find things "charming" every hour, which tired me to death; and every day it was the same dull round of hypocrisy and dissipation. (Edgeworth 2008, 41)

Lady Delacour herself becomes a stimulant for others and thus becomes objectified or 'blocked', a fatal situation in Brunonian terms. The novel criticises the fact that there are lifestyles that run on such forms of external stimulation, be they social or pharmacological.

[6] Opium was a general household item and "[g]oing to the grocer's for opium was often a child's errant" (Berridge 1999, 31). It was used to treat coughs (see 30), colds, sleeplessness, pain, gout, rheumatism, insanity, hysteria (see 32) or "ulcers, bruises, sprains and chilblains" (33). In short: "During the 1800s physicians used opium to treat everything from tuberculosis to diarrhea, and the public employed it as a home remedy" (Mitchell 1988, 559). Opium consumption was common among all classes, even though the ways of consumption also diverge in different social strata, particularly with regard to its "luxuriant" use (Berridge 1999, 49). Among other things, Samuel Taylor Coleridge has consumed the drug in order "to combat the anxiety caused by financial problems" (52), which may also apply to Edgeworth's Lady Delacour. Interestingly, Coleridge was initiated to the drug "through a recommendation of opium in Beddoes' edition of Brown's *Elements of Medicine*" (56). Brunonian medicine is thus the central context for middle-class consumption.

Belinda renders the interdependencies between social structures and drug abuse very explicit. Lady Delacour discloses her knowledge about the psychopharmacology of opium—its "pain-relieving and mood-altering effects" (Berridge 1999, xxi)—to Belinda and thus reveals the way in which individual characters respond to the social requirements of particular social strata. Opium becomes the symptomatic drug that enables both the suppression of pain and a continual performance, which turns opium abuse into a culturally legible marker of particular social structures. Within the context of Brunonianism, the novel thus reflects on whether it is possible for characters to uphold their lives on the basis of justifiable, healthy or unjustifiable, artificial stimulants. In addition, it conflates social forms of exchange within the traditional family with health, thus normalising particular social structures.

Elizabeth Gaskell's *Mary Barton* (1848): Psychopharmacology as Crucial Mediator

Despite the fact that Elizabeth Gaskell states in her Preface to *Mary Barton* that she "know[s] nothing of Political Economy, or the theories of trade" (Gaskell 2008, 6) in order to distance "herself from Harriet Martineau and other professed experts" (Uglow 1993, 192), many readers and critics have focused on this issue (see Dalley 2019, 116–17). The contemporary criticism of Gaskell's novel can be read alongside the marginalisation of the 'political economy' by women in favour of the emergence of a masculine 'economics' (see Vint and Funaki 2017, 63–64). In the *Westminster and Foreign Quarterly Review*, William and Mary Turner Ellis criticise Gaskell for "evidently labouring under the mistaken notion that there is something very difficult, dark and mysterious in political economy. Let her emancipate herself from that error. Very little time will enable an earnest learner to master all the great principles and applications of that science" (Ellis and Ellis 1849, 60). While Gaskell's statements in her Preface can be read as a tactic of self-fashioning by a first-time female author (see Schor 1992, 37), the critical debate shows that the interpretive battles over the rather young genre of the social novel centre around the question whether the social criticism articulated in *Mary Barton* lives up to the tenets of political economy or not and whether the author may be understood as an authoritative voice in the debate. The genre of social realism, in other words, is expected to be 'realistic' in the sense that it conforms to contemporary socio-economic discourses, thus closely aligning the internal, diegetic and external, extradiegetic communication systems of the novel. The publication of a social, or, more precisely, an industrial novel in the tradition of writers such as Charlotte Elizabeth Tonna, Frances Trollope and Harriet Martineau (see Matus 2007, 30) is hence also a political issue; it is not only about romance and the marriage plot, but significantly about the intertwinement of the personal or domestic and the political; hence, the interpretation of *Mary Barton* has long centred on the connection between the plot strands based on John and Mary Barton—politics and romance (see Cooney 2011, 29), but, for both characters, these two are hardly unrelated.

The novel sheds light on the situation of working-class characters in the textile industry in Manchester and their difficulties regarding their options of political participation during the 1840s. The plot strand centring on the seamstress Mary Barton focuses on her options of income and marital choice: her initial romantic attachment to Henry Carson, the son of a wealthy mill owner, and her eventual marriage to a man of her own class, Jem Wilson, with whom she emigrates. While her Chartist father John Barton is guilty of committing a political murder of Henry Carson, it is Jem who is actually charged with it. Mary significantly contributes to clearing up the crime and contracts a brain fever due to the exertion—a very Brunonian idea of Mary exhausting her vital powers in this 'unfeminine' endeavour that also illustrates the continuing cultural impact of Brunonianism in Gaskell's time. After several melodramatic twists and turns, the novel promotes an understanding between different classes and ends on a conciliatory note. Opium features openly in John Barton's attempts to battle hunger and depression, but also implicitly by hints at the context of the Opium Wars that are among the reasons for the economic downturn in the Manchester textile industry. Hence, *Mary Barton* sheds light on the enmeshments of personal lives with Britain's political economy and its colonialist role in the First Opium War.

In recent research, the full scope of Gaskell's social criticism is envisaged in postcolonial approaches with an eye to *Mary Barton*'s transnational scope and the intersection of the plight of the Manchester textile workers with the Opium Wars, which firstly reveal the close interdependencies between drug traffic, the British economy and individual physiologies (see Corley 2003). *Mary Barton*'s participation in the debate on political economy is thus fundamentally intertwined with psychopharmacology. The novel showcases the close ties between the flow of money, the flow of people and the flow of psychoactive substances. While Edgeworth's *Belinda* highlights the connection between individual physiologies and social expectations that are couched in the wider context of economics and medical discourses, Gaskell's *Mary Barton* relates economics more fundamentally to social structures and reveals individual physiologies to be modulated by these structures. Within the social novel, she of course also focuses on a different class than *Belinda*, which is crucial for the discursive approaches to the drug (see Matus 2007, 31–32). Psychopharmacology is the discourse that, as in *Belinda*, reveals that the characters—particularly the working-class characters that are said to be reduced to "the bodily signs of suffering" (Matus 2007, 32)—are very much aware of and reflect on these circumstances. While it is true that "[t]he suffering of the poor is for the most part to be read on starving or diseased bodies or metonymically represented through the state of the home" (Matus 2007, 32), this does not mean that the working-classes are bereft of any interiority and consciousness. Their knowledge may be more implicit, but it is, symptomatically, a psychopharmacological knowledge that reveals their awareness of and responses to their social situation. That this response is encoded as embodied in the novel highlights the severity of suffering that can then be *perceived* by others in sympathy (see Matus 2007, 29–31).

A central *structural* alignment between economy and psychopharmacology is the alteration between 'highs' and 'lows', growth and depression, which is not without wider, ideological implications. As Liam Corley highlights, *Mary Barton* is set

> during the worst periods of economic depression in Manchester in order to depict the suffering that had heretofore gone unmarked by much of the country. The cotton economy of Lancashire experienced seemingly unpredictable cycles of boom and bust throughout the middle of the nineteenth century which caused immense suffering among the workers in the cotton manufacturing industry. (2003, 1)

What is central is that this economic situation is very much an effect of imperial enmeshments, which turns the industrial novel into a 'condition of England in the world' novel (see Corley 2003, 2). Symptomatically, the main action of Gaskell's novel coincides with the First Opium War (1839–42) and reveals the complex economic enmeshments in place in the mid-nineteenth century. "The loss of the China trade was part of the complex factors that led to the unemployment and partial employment depicted in *Mary Barton*", Liam Corley explains, "and it was significant enough to motivate the Manchester merchants to political action" (Corley 2003, 5). The imperial and economic triangle that connects Britain, India and China is central to understand the economic downturn in Manchester and the Lancashire textile industry more widely. The war is fought to secure foreign demand and to expand Britain's own markets, but the tensions initially cause the dire situation in the textile industry due to both India's and China's decreasing demand in English cotton (see Corley 2003, 4).

In *Mary Barton*, "the imperative of guaranteeing foreign demand" (Corley 2003, 3) is central for both the economic realist and the romance plot that crucially and fundamentally intersect in the novel. The Manchester economy is clearly dependent on foreign markets and the 'masters' force their 'hands' to accept meagre wages in order to be able to compete for these markets without communicating the enmeshments of capitalism and imperialism that Gaskell's contemporaries, Karl Marx and Friedrich Engels, explore at the time of *Mary Barton*'s publication (see Marx 1988, 58–59). As the workers take political action and finally go on strike because they and their families are starving due to unemployment and meagre wages, the situation escalates and leads to John Barton killing Henry Carson. While it is true that the novel as a whole seems to favour a sympathetic relationship between workers and manufacturers as a resolution to the situation over any kind of violent revolution (see Matus 2007, 27), it illustrates that the economic situation as it is at that historical moment may lead to such violence and also garners some sympathy and understanding for the workers' course of action.

"Gaskell constructs a vision of domestic social harmony that is necessarily underwritten by the forcible extension and maintenance of empire" (Corley 2003, 3) because the conclusion of the marriage plot depends on the couple's emigration to Canada, where they can lead a wealthier life than at home: "house, - land, - and a good per-centage on the instruments made" are on offer for Mary's spouse, Jem Wilson, in Canada, but not in Manchester (Gaskell 2008, 324). Hence, the marriage plot must be transposed to a colonial space and the reproduction of the couple must occur elsewhere. Mary's class is bereft of viable subject positions on British soil. Colonial expansion is represented as the *conditio sine qua non* for the reproduction of particular strata of the British population, revealing the close enmeshment of colonial and biopolitical discourse formations.

The growth of the population is shown to depend on colonial expansion in a very Malthusian logic that couples the different developments of the increase in the population with the increase in food production. Particularly during the Hungry Forties represented in *Mary Barton*, such Malthusianism renders plausible the articulation of population growth with colonial expansion that provides a more humane solution than starvation. As Thomas Malthus argues:

> IT is an evident truth, that whatever may be the rate of increase in the means of subsistence, the increase of population must be limited by it, at least after the food has once been divided into the smallest shares that will support life. All the children born beyond what would be required to keep up the population to this level, must necessarily perish, unless room be made for them by the deaths of grown persons. (Malthus 1809, II 359)

Mary Barton's solution to this problem by way of emigration thus represents rather lenient consequences to the situation in 1840s Manchester for the protagonist, but the novel is equally explicit about the many people starving or "clemming" to death (Gaskell 2008, 60). Foucault's biopolitical formula—"'make' live and 'let' die" (2004, 241)—in the sense of a management of populations is relentlessly described in Gaskell's industrial novel, and opium is the drug that is tied to the social situation described. The physiology of hunger and the known drug effects of opium are crucially intertwined and illustrate the entanglement of Britain's domestic economy with colonial trade on a physiological level.

Mary's enmeshment in these global interdependencies is thrown into relief when she needs to occupy the bed of a sailor to recover from the brain fever she contracts after her courageous intervention in Jem Wilson's trial. Jem is charged for the murder that Mary's father has committed and Mary seems to be the only agent who can exonerate Jem without incriminating her father; she is able to produce a witness to confirm Jem's alibi, an undertaking that proves too much a strain for her; she succumbs to illness after an overstimulation of her brain and her physical powers through this endeavour. After the trial, Mary resides with the Sturgis family to recover: "Mrs. Sturgis led her into a little room redolent of the sea and foreign lands. There was a small bed for one son, bound for China; and a hammock slung above for another, who was now tossing in the Baltic" (Gaskell 2008, 273). Thus, Mary "will rave and recuperate in a space created by the absence of a sailor 'bound for

China'" (Corley 2003, 3). The creation of spaces in the novel is thus part of its politics. Which (subject) positions are shown to be viable and which spaces can a character rightfully occupy? Mary can occupy the sailor's bed only because he, as the novel insinuates, leaves to fight in the First Opium War. His room, however, is still reminiscent of "the sea and foreign lands" (Gaskell 2008, 273), and neither his bodily presence—perhaps still to be felt by the indentures in his mattress—nor his journey that requires Britain's sea power to reach other countries such as China fully recede from the characters' and the readers' consciousness. Mary can temporarily fill the sailor's position and her recuperation depends on that position; in other words, Mary's health is dependent on foreign markets; without these markets, there would not be any 'room' for her. The long debate about the novel's allegedly incommensurable double plot must hence be changed to the utter interdependence of these plots—the plight of John Barton and the plight of Mary Barton both depend on foreign markets and Britain's imperial global relations and both their plots are connected to opium and the First Opium War. This does not only go for the marriage plot (see Lee 2019, 515), but also for Mary's situation as a worker who depends on the textile industry as a seamstress. In that capacity, she provides the family income, which soon flows into John Barton's opium addiction: "She could not stay often to make purchases of food, but gave up the money at once to her father's eager clutch; sometimes prompted by a savage hunger it is true, but more frequently by a craving for opium" (Gaskell 2008, 125).

The situation of the Bartons is circumspectly described. After bereavement and unemployment, John Barton turns to opium to face both his mental suffering and physical deprivation. Soon, his drug consumption becomes habitual, "smoking, or chewing opium. Oh, how Mary loathed that smell!" (105). Through the focalisation on Mary, the narrator illustrates the consequences of Barton's dependence for his social relations: "He wanted it to relieve him from the terrible depression its absence occasioned. A large lump seemed only to bring him into a natural state, or what had been his natural state formerly" (110)—and offers a sympathetic perspective with which the readers are invited to align themselves:

> But before you blame too harshly this use, or rather abuse, try a hopeless life, with daily cravings of the body for food. Would you not be glad to forget life, and its burdens? And opium gives forgetfulness for a time. It is

true they who thus purchase it pay dearly for their oblivion; but can you expect the uneducated to count the cost of their whistle? (149–50)

Besides this pharmacological explanation—opium staves off any feeling of hunger and alleviates pain, both physical and mental—this comment illustrates the social circumstances that may lead to habituation and also draws attention to the benefit-cost-calculation that may render oblivion more attractive than consciously suffering one's fate, and with this calculation, also proves to be a rational choice in the context of one of the novel's leading discourses, political economy. As the novel has shown so far, John Barton tried different levers of individual political agency—he is a Chartist, a member of a trade union, and he spearheaded the commission to inform Parliament about the workers' plight (Gaskell 2008, 23; 25), but because he realises that all his attempts at agency are curtailed, he succumbs to opium. The narrator also provides the economic reasons for opium abuse: "Many a penny that would have gone little way enough in oatmeal or potatoes, bought opium to still the hungry little ones, and make them forget their uneasiness in heavy troubled sleep" (Gaskell 2008, 53). Opium is cheaper than food and hence may alleviate suffering more lastingly than other purchases that may seem more rational to the middle-class observer.

Mary Barton thus also belongs under the umbrella term of addiction literature, an important addition to its classification as social and industrial novel. The focus on industrialisation reveals that capitalism and imperialism intersect in British industries such as the textile industry in Manchester; the focus on society illustrates the impact of these interdependencies on each individual member of society, stratified along the lines of class, race and gender; the focus on drug consumption reflects both the imperial enmeshments and the effects of the social stratification on individual physiologies. Opium in particular is indicative of these intersections. Following its trade routes—from India to China and, mainly through revenues, back to Britain—and the practices of its consumption—eating, smoking, drinking solutions such as laudanum and the corresponding spaces of such consumption—one can easily establish a sociopolitical map of Britain in the world that illustrates the flows of people, goods and capital and thus the physical, economic and political desires these streams cater to. *Mary Barton* is aptly described as a condition-of-England-in-the-world novel; it is the flows of opium in blood streams that serve as an indicator of the interconnections between individual and political dependencies.

Psychopharmacology not only mediates the position of England in a global context, but also serves as a marker of a transition in medical knowledge and practices. The novel clearly illustrates the shift in authority from a "vernacular medical tradition" with "its deep epistemological links with natural history" (King 2003, 256) to "regular medicine" (259). From the very beginning of *Mary Barton*, there is a rural idyll that is clearly marked as *past*, constructed as an idealised, spacious and natural setting that is intricately connected to the vernacular medical tradition. The garden was the druggist's shop and it was mainly women who collected herbalist knowledge. Regular medicine, however, is coded as much less beneficial and is associated with masculinity in the novel. Alice Wilson embodies the altruistic herbalist who cares for her community of neighbours (see King 2003, 255–56):

> She had been out all day in the fields, gathering wild herbs for drinks and medicine, for in addition to her invaluable qualities as a sick nurse and her worldly occupations as a washerwoman, she added a considerable knowledge of hedge and field simples, and on fine days, when no more profitable occupation offered itself, she used to ramble off into lanes and meadows as far as her legs could carry her. (Gaskell 2008, 17)

What is more, she was preparing medicine for Mary Barton, the young Mary Barton's mother, who dies in childbirth. The doctor arrives too late to provide any help. Male druggists and doctors are characterised by economic and professionalised discourses rather than through discourses of caring. While the herbalist is there for everyone, medical professionals seem to have a clear class bias. This is illustrated by the failure of a druggist to help save John Barton's friend Davenport:

> The druggist (whose smooth manners seemed to have been salved over with his own spermaceti) listened attentively to Barton's description of Davenport's illness; concluded it was typhus fever, very prevalent in that neighbourhood; and proceeded to make up a bottle of medicine, sweet spirits of nitre, or some such innocent potion, very good for slight colds, but utterly powerless to stop, for an instant, the raging fever of the poor man it was intended to relieve. (Gaskell 2008, 57–58)

The main aim here seems pecuniary; the druggist sells medicine he knows to be useless in the case of typhus fever. John Barton is gullible enough and has sufficient faith in the druggist's medical authority not to question this treatment, but the plot reveals the futility of the whole undertaking when Davenport dies.

These different forms of medical knowledge are pitted against one another in the novel and in terms of psychopharmacology; the vernacular herbalist tradition is definitely to be preferred over regular medicines such as opium. *Mary Barton*'s psychopharmacology thus co-modulates the adhesive powers of society and the condition of the possibility of a human community regardless of markers of race, gender and class. The (mental) health of the social community is a function of the kinds of medicines, vernacular or regular, the ways in which they are administered, and the individual blood streams through which they flow. While the ideology of the novel may still be taxonomic and visual (King 2003, 261–62), it nevertheless offers a visionary remedy for social disintegration by salvaging a vanishing past, very much defined by Alice's herbalist knowledge.

Opium in Women's Writing

For early nineteenth-century women poets, Hannah Cowles Freeman argues, "opium assists in a politics of cooperation and compromise by allowing them temporary escape from an unsatisfactory gender system which left many women feeling confined and limited" (2012, 15). Even though this function is still visible in early nineteenth-century women's fiction, particularly in Edgeworth's *Belinda*, women writers not only illustrate the close articulation between gendered domesticity and opium consumption, but also couch these concerns in wider contexts such as domestic economy and emerging discourses on psychopharmacology. For the upper-class Lady Delacour, this connection is very much part of her personal financial situation, while the lives of the working-class characters represented in *Mary Barton* are shown to be delicately dependent on a global economy. Both novels emphasise free exchange. For Lady Delacour, the obsessive and habitualised repetition of behaviour to play—and ossify—the role as a woman of society is finally supplanted by a free exchange in a lively community of friends and family, albeit not without the obligation to closely monitor behaviour patterns in order to prevent the emergence of potentially harmful and excessive habits; Brunonian medical discourses on the moderation of stimuli pave the way for a biopolitical risk management of health. As Belinda serves as the 'medicine' that brings about Lady Delacour's turn to 'health' by way of her communication with her patient, novel readers are included in a process of doing psychopharmacology; the very performance of reading may induce a similar turn to health in the moderate reader. In *Mary Barton*, such free exchange already resembles an imperative to free trade; the novel reveals that, in terms of domestic economy, Britain is part of a transnational system of economic exchange that depends on foreign markets to thrive—from a global economic downturn to a very personal level of 'health', marriage and procreation. Since the latter are all biopolitical concerns, the novel illustrates the close intertwinements of domestic and foreign economy, colonialism and biopolitics.

Harriet Martineau, whom Edgeworth surmised to be the author of *Mary Barton*, provides the political economy for the notion of free trade that comes to designate the ideology of Gaskell's novel in particular, albeit with the requirement for free trade to be held in check by individual sympathy. In her "Moral of Many Fables" in *Illustrations of Political Economy*, Martineau argues:

> In order to see how extensively and how effectually governments have interfered to pervert the natural distribution of the gifts of Providence, it would be necessary to review almost the whole list of spontaneous and artificial productions; for there are few or none whose spread has not been arbitrarily stopped in one direction or another. What Great Britain alone, – the most enlightened of commercial countries, – has done in damming up the streams of human enjoyment, is fearful to think of. (Martineau 1834, 108)

Hailing both 'nature' ("the natural distribution") and religious discourses ("the gifts of Providence"), Martineau naturalises free trade (108). 'Damming up the streams of human enjoyment' is a detrimental economic tactic that finds an analogy in the Brunonian concept of health in *Belinda*—health is free and open (narrative) exchange without blockages or 'damming'—and the plea for a free exchange of goods in foreign markets in *Mary Barton*.

Beginning her *Illustrations* five years prior to the First Opium War, Martineau considers the East India Company's monopoly in the opium trade detrimental and finds the solution in free trade:

> Whatever may be thought of the indulgence of opium in this country, no one objects to its being used by the Hindoo and the Chinese as a stimulus appropriate to the climate in which they dwell. If we had allowed things to take their *natural* course, Persian husbandmen would have tended their vast poppy-fields, season by season, guarding that delicate plant from the injuries of insects, and sheltering it from unfavourable winds, while the Chinese and the Hindoos would have been busy preparing commodities to exchange with the Persian, and all would have been made rich enough by their *traffic to keep British merchant-ships continually going and coming* to supply their wants. [...] By *universal and free exchange*, – that is, by each person being permitted to exchange what he wants least for what he wants most, – an absolutely *perfect system of economy* of resources is established; *the whole world being included in the arrangement*. (109–10; 113; emphasis added)

For Martineau, free trade is universal and global; it is idealised as *perpetuum mobile* for the flow of goods, money and people. Since free trade is considered to be 'natural', it must be granted an organic expansion, it cannot be forced. Colonialism thus seems to be a natural factor of economic growth and development: "Colonies are advantageous to the mother-country as affording places of settlement for her emigrating members, and opening markets" (1834, 115). This seems to be the backdrop for Gaskell's plot in *Mary Barton*, where people who are 'least wanted'—in Manchester, for instance—can migrate to places where they are most wanted, such as Toronto.[7]

Whereas this seems to be the ideological normalisation of the interconnection between capitalism and imperialism, Martineau also articulates an intercultural approach to the ensuing economic global ties that betrays a considerable hermeneutic sensitivity. In the case of the international relations between Britain and China, she is very aware of the potentially problematic confrontation of auto-images and hetero-images: "Martineau's earliest writing about China, 'The English in China. A Tale' was published in 1832, the same career-launching period that produced *Illustrations of Political Economy*. [...] This tale provides surprising insights into western Orientalism, eastern Occidentalism, and the ethnocentrism of both" (Logan 2016, 167). She considers a failure to understand such

[7] These are places that also mark destinations for people trying to escape from Black '47, which again emphasises the dire economic situation on the British Isles during the Hungry Forties.

hermeneutic challenges as part and parcel of the Opium Wars: "The First Opium War, a direct result of opening China to free trade, constitutes an episode in imperial history that casts British aggression in China as indefensible, 'a national disgrace' [...] stemming from cultural arrogance, linguistic ignorance, and a refusal to honor Chinese socio-political mores" (Logan 2016, 169). She is very aware of the effects of epistemic violence in these international relations when she argues that "the hope of being wiser in future must compensate to us, as a nation, for the mortification and shame of discovering what prodigious mistake we have made, and what terrible things we have done, by bringing our English notion to bear on affairs which they could in no way explain" (Martineau 1858, 372–73). Hence, economic expansion must be dovetailed by intercultural knowledge.

Both Edgeworth's *Belinda* and Gaskell's *Mary Barton* employ psychopharmacology to illustrate the close interconnection between ideologies of free trade and imperialism, individual bodies and a socio-cultural stratification along the lines of race, gender and class. Individual desires and bodily cravings come to be legible as cultural practices and markers that illustrate the way in which individual viability has become a function of global economic enmeshments. Early nineteenth-century fiction by women writers is thus clearly not restricted to the domestic but reveals the domestic as always already intertwined with transnational political economy. Opium serves as a shorthand for such enmeshment.

References

Angell, Marcia. 2018. Opioid Nation. *The New York Review* 56 (19): 56–58.

Berridge, Virginia. 1978. Victorian Opium Eating: Responses to Opiate Use in Nineteenth-Century England. *Victorian Studies* 21: 437–51.

———. 1999. *Opium and the People: Opiate Use and Drug Control Policy in Nineteenth and Early Twentieth Century England*. London: Free Association.

Brown, John. 1795. *The Elements of Medicine*, 2 vols. London: J. Johnson.

Budge, Gavin. 2012. *Romanticism, Medicine and the Natural Supernatural: Transcendent Vision and Bodily Spectres, 1789–1852*. New York: Palgrave Macmillan.

Bynum, W.F., and Roy Porter. 1988. Introduction. In *Brunonianism in Britain and Europe*, ed. W.F. Bynum and Roy Porter, ix–x. London: Wellcome Institute.

Collins, Wilkie. 2008. *The Moonstone*, ed. John Sutherland. Oxford: Oxford University Press.

Colman, Adam. 2019. *Drugs and the Addiction Aesthetic in Nineteenth-Century Literature*. Cham: Palgrave Macmillan.

Cooney, Brian. 2011. Violence, Terror, and the Transformation of Genre in *Mary Barton*. In *Victorian Transformations: Genre, Nationalism and Desire in Nineteenth-Century Literature*, ed. Bianca Tredennick, 29–44. Farnham and Burlington: Ashgate.

Corley, Liam. 2003. The Imperial Addiction of *Mary Barton*. *The Gaskell Society Journal* 17: 1–11.

Dalley, Lana L. 2019. Women's Writing and the Mainstreaming of Political Economy. In *The Routledge Companion to Literature and Economics*, ed. Matt Seybold and Michelle Chihara, 114–22. London and New York: Routledge.

Edgeworth, Maria. 2008. *Belinda*, ed. Kathryn J. Kirkpatrick. Oxford: Oxford University Press.

Ellis, William, and Mary Turner Ellis. 1849. Mary Barton. *Westminster and Foreign Quarterly Review* 51: 48–63.

Foucault, Michel. 1982. The Subject and Power. In *Michel Foucault: Beyond Structuralism and Hermeneutics*, ed. H. Dreyfus and P. Rabinow, 208–26. Chicago: University of Chicago Press.

———. 2004. *'Society Must Be Defended': Lectures at the Collège de France, 1975–76*, trans. David Macey. London: Penguin.

Freeman, Hannah Cowles. 2012. Opium Use and Romantic Women's Poetry. *South Central Review* 29 (1–2): 1–20.

Gaskell, Elizabeth. 2008. *Mary Barton: Authoritative Text, Contexts, Criticism*, ed. Thomas Recchio. London and New York: Norton.

Haiven, Max. 2018. Our Opium Wars: The Ghosts of Empire in the Prescription Opioid Nightmare. *Third Text* 32 (5–6): 662–69.

King, Amy Mae. 2003. Taxonomical Cures: The Politics of Natural History and Herbalist Medicine in Elizabeth Gaskell's *Mary Barton*. In *Romantic Science: The Literary Forms of Natural History*, ed. Noah Heringman, 255–70. Albany: The State University of New York Press.

Larson, Leah. 2009. Breast Cancer and the 'Unnatural' Woman in Edgeworth's *Belinda*. *The Explicator* 67 (3): 195–98.

Lee, Michael Parrish. 2019. Gaskell's Food Plots and the Biopolitics of the Industrial Novel. *Victorian Literature and Culture* 47 (3): 511–39.

Logan, Deborah A. 2016. *Harriet Martineau, Victorian Imperialism, and the Civilizing Mission*. London and New York: Routledge.

Malthus, Thomas Robert. 1809. *An Essay on the Principle of Population*, 2 vols. Washington: Robert Chew Weightman.

Mansel, Henry. 1863. Sensation Novels. *Quarterly Review* 113: 482–514.

Martineau, Harriet. 1832. The English in China. *Tait's Edinburgh Magazine* 1: 667–83.

———. 1834. *Illustrations of Political Economy, Vol. 9*. "The Farrers of Budge Row". "The Moral of Many Fables". London: Charles Fox, Paternoster Row.
———. 1858. China: Past and Present. *Westminster Review* 69: 370–401.
Matus, Jill. 2007. *Mary Barton and North and South. The Cambridge Companion to Elizabeth Gaskell*. Cambridge: Cambridge University Press, 27–45.
McAllister, Marie. 2015. Ungovernable Propensities: Belinda and the Idea of Addiction. *The Age of Johnson* 23: 301–31.
Mitchell, Dennis J. 1988. Opium. In *Victorian Britain: An Encyclopedia*, ed. Sally Mitchell, 559–60. New York: Garland.
Murison, Justine S. 2015. Obeah and Its Others: Buffered Selves in the Era of Tropical Medicine. *Atlantic Studies* 12 (2): 144–59.
Peters, Dolores. 1981. The British Medical Response to Opiate Addiction in the Nineteenth Century. *Journal of the History of Medicine and Allied Sciences* XXXVI (4): 455–488.
Schor, Hilary M. 1992. *Scheherazade in the Marketplace: Elizabeth Gaskell and the Victorian Novel*. Oxford: Oxford University Press.
Siegel, Shepard. 1983. Wilkie Collins: Victorian Novelist as Psychopharmacologist. *Journal of the History of Medicine and Allied Sciences* 38 (2): 161–75.
Sutherland, John. 2008. Introduction. In *Wilkie Collins: The Moonstone*, ed. John Sutherland, vii–xxix. Oxford: Oxford University Press.
Teynor Donatini, Hilary. 2011. Moral and Medical Diagnosis in Maria Edgeworth's *Belinda*. *The Eighteenth-Century Novel* 8: 247–69.
Uglow, Jennifer. 1993. *Elizabeth Gaskell: A Habit of Stories*. London and Boston: Faber and Faber.
Vint, John, and Keiko Funaki. 2017. Harriet Martineau and Classical Political Economy. In *Harriet Martineau and the Birth of Disciplines: Nineteenth-Century Intellectual Powerhouse*, ed. Valerie Sanders and Gaby Weiner, 49–68. London and New York: Routledge.
Zieger, Susan. 2011. Opium, Alcohol, and Tobacco: The Substances of Memory in *The Moonstone*. In *A Companion to Sensation Fiction*, ed. Pamela K. Gilbert, 208–19. Malden, Oxford and Chichester: Wiley-Blackwell.

CHAPTER 10

The Indeterminate Pharmacology of Absinthe in Nineteenth-Century Literature and Beyond

Vanessa Herrmann

The story of absinthe is riddled with legends, assumptions and misconceptions as well as wondrous tales of inspiration sweeping over artists. But apart from that, it is also the story of addiction. This chapter locates the history of absinthe within the discourse of addiction to alcohol. In the late 1990s, most European countries lifted the ban on absinthe, and in 2007, the USA followed. In this period, popular interest in absinthe was renewed, and the substance was sold as a popular party drink. In addition, scientific research on absinthe and its supposedly addictive ingredient thujone was undertaken. This chapter sets out to establish not only the connections between alcoholism and absinthism but also the indeterminate pharmacology of absinthe itself, thereby questioning the prevalent prejudices surrounding the drink. In nineteenth-century literary texts, the word *dangerous* is used frequently when describing the beverage; hence, sources using said vocabulary have to be dealt with critically. The history

V. Herrmann (✉)
University of Bremen, Bremen, Germany
e-mail: vanessa.herrmann@uni-bremen.de

N. Roxburgh and J. S. Henke (eds.), *Psychopharmacology in British Literature and Culture, 1780–1900*, Palgrave Studies in Literature, Science and Medicine,
https://doi.org/10.1007/978-3-030-53598-8_10

of the beverage and the discussion of its effects are closely related to that of the temperance movement and to the treatment of alcoholism as a disease. Various medical approaches of the nineteenth century are therefore considered here.

Research has shown that it is thujone—inherent to the wormwood plant—that supposedly causes side effects such as "convulsions, hallucinations, and insomnia" (Judžentienė 2016, 855). These effects were subsumed under the name *absinthism*, an illness believed to be derived from excessive consumption of the beverage that gave it its name. However, recent research does not agree on the drug effects of thujone; this particular issue will be at the centre of this chapter, as it influences our reading of nineteenth-century literature and the question of alcohol-induced addiction. Neither contemporary nor previous knowledge about drugs produced a consensus on the effects of thujone, and this ambivalent attitude towards absinthe is reflected in nineteenth-century literature. Texts by Christina Rossetti and Robert Hichens, used as case studies in this essay, reveal such ambivalent attitudes towards absinthe, whose effects are shown in both cases to be indeterminate.

Absinthe and Thujone: A Short History

The use of wormwood can be traced back to Ancient Egypt, as a medicinal document called the Ebers Papyrus dated around 1550 BC reveals (Judžentienė 2016, 850). It was used from then onwards to treat various illnesses, such as insomnia, headache, indigestion, fatigue or to heal wounds and insect bites (850). For centuries, the plant and preparations made from and with it were part of medical practice. As Jad Adams notes, it was often connected to the treatment of illnesses or problems relating to sexual intercourse or the reproductive organs (2004, 17). It is still used today as an essential oil and "as a fragrance component in soaps, cosmetic and perfumes" (Judžentienė 2016, 850). Despite its known aphrodisiacal effects, it would be used for a wider variety of purposes later on.

Wormwood continued to be used as medicine until well into the nineteenth century. The alcoholic beverage named absinthe was invented in the late eighteenth century, but it became popular after French troops used it as an antiseptic and as a medicine to treat indigestion during the Algerian Wars (1844–1847) (Lachenmeier et al. 2006, 366). They brought the beverage back to France and it became a favourite with bourgeois Parisians. None of the histories of absinthe mention alcoholism or addiction at this time. The drink was not associated with negative effects; in fact, imbibing was often seen as a patriotic gesture (Adams 2004, 9). The popularity of the spirit coincided with and was probably facilitated by the liberation of licensing laws under the reign of the Second Empire (1852–1870); more people were allowed to sell alcoholic beverages, and the popular drink absinthe was one of them (Adams 2004, 47). But its popularity also had a downside: cheap 'copies' of the original drink surfaced using less pure alcohol (Lachenmeier et al. 2010, 33). Absinthe became a mass phenomenon as various parts of nineteenth-century society partook in its consumption.

These developments can be found mainly in France, but the absinthe craze swept over the Channel to Britain and especially London. One of the most famous names associated with the drink was (and still is) Oscar Wilde. As part of the Decadence movement, absinthe soon became the signature drink of the participants. According to Jad Adams, "[f]or English writers and artists of the 1880s and 1890s, France represented everything new and challenging in art, all symbolised by the diabolical green drink. Artists became not only pro-French, but anti-English" (2004, 8). The growing sentiment towards the French was met with suspicion on the part of the bourgeois nationalists, and they found the embodiment of their suspicions in absinthe, the popular French drink.

The decline of absinthe use began with the publicised illness of absinthism. This disease was supposedly caused by the consumption of large quantities of absinthe. The symptoms diagnosed in the nineteenth century included: visual and auditory hallucinations, epilepsy, mental defects and a heightened propensity for suicide (Hein et al. 2001, 2717). Lachenmeier writes that in the past "convulsions, blindness, hallucinations, and mental deterioration" (2006, 312) were the diseases caused by the customary consumption of absinthe. He goes on to explain that "[i]n the advanced state, signs of degeneration could be observed, which could cause convulsions that even resulted in death" (312). It is important to note that this illness was diagnosed using problematic words such as *degeneration* (which evokes eugenics) 150 years ago, and that twenty-first-century research suggests another interpretation of the substance's drug effects.

Nineteenth-Century Views on Absinthe and Alcohol Consumption

As absinthe is an alcoholic beverage—and both past knowledge and modern knowledge about drugs treat it as such—many parallels can be found to the terminology and treatment of alcohol addiction. The manner in which physicians of the late nineteenth century dealt with addiction in general is telling: They treated it as an illness, a disease of the mind that not only corrupted one's morals but was also hereditary (Foxcroft 2007, 119). Alcoholism is no exception here: "[P]hysicians held that alcoholism was a disease – characterised by the drinker's inability to control his or her intake of alcohol" (Warhol quoted in Brodie Farrell and Redfield 2002, 100). It is this attitude which turned the addict into a social pariah. The lack of will to resist alcohol was also connected to a moral weakness (Foxcroft 2007, 119). Susan Zieger points out that a popular nineteenth-century response to alcoholism, which still occurs in contemporary discourse on the subject, was the following: "[A]ddiction may be a disease, but it is one that can be overcome by the sufferer's will, conceived of primarily as the power of mind over matter" (2008, 2). Both Louise Foxcroft and Zieger reveal the attitude towards addiction as a disease of the morally weak. Thus, the addicts had to be taken care of professionally by imprisoning them in asylum-like institutions; the first one of these inebriate asylums was founded in 1857 (Warhol quoted in Brodie Farrell and Redfield 2002, 100). Treatment varied from complete prohibition to administering small dosages of alcohol; alcoholism was even treated by giving wine to the addict, a cure mainly used in France (where wine and beer were seen more as socially acceptable drinks).

The differentiation between the two kinds of alcohol and their drug effects shows the complexity of the discussion surrounding absinthe itself. It sheds light on the beginnings of the difficult analysis of what absinthe is and can do. For instance, the French word *alcool* refers to spirits but not to wine or beer (Adams 2004, 9). Therefore, spirits like absinthe were condemned and perceived as a bad influence on people whereas wine was perceived as a healthy drink. According to Adams, "[it] was widely believed that the problem with alcohol was not the quantity consumed, but the quality" (2004, 9). Adams goes on to point out that even "temperance organisations maintained that a litre of wine a day was a reasonable dose for a healthy man" (9). This development was facilitated by the discourse of good/bad and pure/impure alcohol; the vocabulary alone shows an over-simplification of the discussion and the rhetoric of religion or morality rather than science.

Early eighteenth-century medical practitioners were also concerned with the effects of overconsumption of alcohol. Thomas Trotter, for example, stated that alcohol "disturbs, or operates on the animal functions, that new affections of mind, latent, or unknown before, are produced; and the drunkard appears to act the part of a man of deranged intellect, and altogether foreign to the usual tenor his sober reflections" (1810, 12). A distinction between good and bad alcohol can be found in John Brown's late eighteenth-century theory of neuropathology, which implied that health and disease form a proportional relationship (Vickers 1997, 66). The implication here is that both health and sickness are variations of the state of excitement (Youngquist 1999, 155). At the core of Brown's theory, we can find two states of the body that need to be treated: "If there is an excess, the condition is called *sthenia*; the opposite deficiency Brown named *asthenia*" (Risse 1988, 48; italics by Risse). Brown shared the belief of many medical practitioners that certain kinds of alcohol have beneficial characteristics and can be used for treating particular diseases.

In this context, the analysis of absinthe was also subject to this vocabulary. Its ingredient wormwood set it apart from the other alcoholic beverages, but as its effects on the body had not been determined precisely enough, medical practitioners were confused as to how to deal with it in the context of alcohol-induced addiction. It has to be noted at this point that thujone, the decisive ingredient in absinthe, was only discovered in the year 1894 by Friedrich Wilhelm Semmler at the University of Greifswald (see Semmler 1894). Hence, most of the discussion of absinthe in the nineteenth century was based on the analysis of known drug effects of alcohol—but not on its chemical components.

The leading researcher on alcoholism in the nineteenth century was Dr Valentin Magnan (Prestwich 1979, 305). Patricia Prestwich states that Magnan was actually one of the first practitioners to look for psychological causes of alcoholism (305). However, he also identified absinthe as the cause for many "medical and psychological troubles not associated with the high consumption of alcohol", as the "essences of absinth [...] attacked the nerves, particularly of head and neck, before the alcohol had time to act" (306). This attack then supposedly led to convulsions, insanity or death. Magnan also saw these effects as irreversible and hereditary. As his theories were spread within the scientific community and also in the journal *The Lancet* (one of the oldest medical publications to date), people soon associated these illnesses with the consumption of absinthe.

Apart from looking into the effects of absinthe on the body, medical discourse emphasised preventing the spread of drinking. According to Prestwich, little time was invested in curing or treating the victims of alcoholism (1979, 307). Absinthe was deemed the reason for many problems gripping French society: increasing rates of crime and violence, decreasing rate of births, decreasing number of men capable of becoming soldiers. The abolition of absinthe seemed to be the easy cure-all for these complex problems (Prestwich 1979, 307–08).

In the nineteenth century, however, these mechanisms were yet to be understood, so finding a single cause seemed to be a good way to improve the overall situation. Similar to the discussion of the effects of opium addiction, the aspect of heredity also played an important part in the condemnation of absinthe. Foxcroft sums up the prevalent attitude towards addiction in general:

> The formation of the disease entity of addiction was based on the question of will, on voluntary and involuntary reflexes which were a function of the nervous system. Nervous diseases or disturbances were themselves thought to have their origins in heredity, and, if sufficiently severe, were diagnosed as a form of insanity. Where addiction was concerned this was a 'moral insanity', a hybrid disease, a physiological affliction with behavioural symptoms that were open to judgement. (Foxcroft 2007, 119)

Foxcroft here goes along the lines of discussion of Prestwich and Zieger: they point out the role of will power in the nineteenth-century understanding of addiction. It was seen as both a nervous condition and a social category. Those who were afflicted with alcoholism and/or absinthism were considered impressionable and thus deemed weak members of society. In this context, it was absinthe's ingredient wormwood that played a decisive role. According to Prestwich, if "alcohol was a poison that led to madness and death, then absinth provided a particularly horrible form of lunacy and a faster death" (1979, 308). Thus, addiction became a social category: addicts were described as mentally ill and thus treated in hospitals like Magnan's.

The recurring notion of insanity as a result of addiction was one of the central arguments against drugs or alcohol. In France, the idea of the mental decline of the following generation took hold after the numbers of alcohol addicts rose. Magnan was one of those physicians to promote the theory that absinthe was the main reason for this 'degeneration' (Prestwich 1979, 306). Foxcroft further points out that "the question of heredity influenced many other areas of social concern during the second half of the nineteenth century [...]. Addiction was seen as a form of inherited profligacy and a predisposition of insanity, a chain of cause and effect stretching across generations" (Pick quoted in Foxcroft 2007, 125). Foxcroft's findings reveal that the discussion surrounding addictive substances was far from what modern research has shown.

Magnan not only advanced the field of psychology in Europe in the nineteenth century, but he was also one of the strongest opponents of absinthe. Prestwich states that Magnan was a "respected authority on insanity (particularly alcoholic insanity)" (Prestwich 1979, 305). Despite his influence, Magnan's research was and still is viewed critically. According to Prestwich,

> his effectiveness was limited by his preoccupation with the so-called 'degeneration' of the French race and by his unbending conviction that alcohol was the prime cause of most mental disorders. His interpretations of the problem of alcoholism, and particularly absinthism, were consequently simplistic and alarmist, and he attributed to alcohol many types of nervous illnesses that today are considered to have more diverse origins. (1979, 305–06)

However, the publication of Magnan's article "De l'alcoolisme" (1874) firmly established the connections between absinthism, insanity and alcoholism (Prestwich 1994, 321). In a paper delivered at the First International Eugenics Congress held in London in 1912, Magnan reasserted the results of his work in the asylum, providing important background information regarding the general description of alcohol. Apart from describing absinthism in detail, he stated that "[of] 1000 children of alcoholics, about one-third disappears at birth or in the first two or three years; amongst the survivors are counted many idiots, epileptics, and a large number of degenerates destitute of moral sense, instinctively perverted, impulsive, abnormal, miserable victims of their parents' alcoholism" (Magnan and Fillassier 1912, 8). The more alcohol is consumed, the higher the chances of a so-called 'degenerate' next generation, he argued. Absinthe was identified as the worst kind of alcohol possible. Magnan's findings do not, however, distinguish between alcohol in general and absinthe in particular. These results—which fed into the coming century's dark history of eugenics—played into the hands of the temperance movement that fought to ban absinthe and other alcohols. Prestwich mentions that at meetings of some of these groups "innocent guinea pigs and rabbits were injected with doses of pure essence of absinth and quickly died a horrible, convulsive death" (1979, 306). Naturally, these 'experiments' were not conducted by physicians, chemists or the like, but by those leading the groups. One has to point out, however, that these demonstrations were based on actual trials conducted by Magnan himself. He "used clinical observations of dogs treated with absinthe oil to infer, as in humans, the presence of hallucinations" (Lachenmeier et al. 2010, 34). Lachenmeier states that Magnan's results were challenged in both France and Great Britain, but to no avail (34).

Twenty-First-Century Research on Absinthe

When turning to more recent research on absinthe and its effects, the debate has not yet come to an end. Scientists still do not agree on the effect of thujone, the active ingredient in wormwood. On the one hand, there are researchers who believe that thujone has similar effects to THC in cannabis, amongst them Jiří Patočka and Bohumil Plucar (2003) from the Czech Republic and Jakob Hein, Lars Lobbedey and Klaus-Jürgen Neumärker from Germany (2001). The opponent to this theory (and to disease category of absinthism) is Dirk W. Lachenmeier, with his several research teams (2006, 2007, 2010).

Hein and his team support Magnan's earlier idea that the consumption of large quantities of absinthe can cause severe psychological and physiological illnesses. Hein postulates that the specific neurotoxic effect of the beverage stems from its thujone concentration (2001, 2716). He also points out that thujone—as a toxic substance—has a similar molecular structure to THC (2001, 2718); he refers here to the findings of J. del Castillo and others in an article titled "Marijuana, Absinthe and the Central Nervous System" from 1974. The conclusion drawn from this similarity supports Hein's theory of thujone being a hallucinogenic drug. Hein's own research replicated Magnan's experiments on dogs; he too injected high doses of thujone into rats, which then suffered from convulsions of first tonic and neuroclonic character (Hein et al. 2001, 2718). For Hein, this proves the devastating consequences of imbibing vast amounts of absinthe. Patočka and Plucar confirm these findings, although they concede that "thujone's mechanism of action is unknown" (2003, 201). They conclude that "absinthe drinking caused both acute and chronic toxicity" (203). Furthermore, it is "addictive and causes personality changes because of cortical damage which can lead to recurrent and sporadic seizures resembling epilepsy" (204). Both groups of scientists assert Magnan's ideas when it comes to the negative effects of absinthe and the illness identified by him.

On the opposing side, Lachenmeier and his team(s) showed in several analyses that "scientific literature contains no proof that historic absinthes may have contained thujone in concentrations able to produce toxic effects" (2007, 306). They analysed the chemical composition of thujone as well as the thujone level in historic absinthes from the pre-ban era and contemporary versions of the beverage. In their essay from 2010, Lachenmeier and his team stated the following result:

> The total thujone content in 13 analyzed pre-ban samples was found to range between 0.5 and 48.3 mg/L, while the average thujone content of 25.4 ± 20.3 mg/L fell within the modern *Codex Alimentarius*/European Union (EU) limit of 35 mg/L (which was derived from no-effect levels in animal experiments with an additional safety factor). [...] From an analytical viewpoint, these findings effectively exonerate absinthe of its sinister reputation. (35)

Their findings contradict the results of the other two teams of researchers. In two preceding articles, Lachenmeier also points out that people in general (and his colleagues in particular) "should focus more on real analytical and toxicological facts instead of resuming all the factoidal myths and legends of former years" (2006, 374). In his publication from 2007, Lachenmeier openly attacks Hein and his team and sets out to disprove their findings. These direct attacks show not only a strong conviction that his own results are correct, but it also reveals that the tension between these two attitudes has not weakened even after more than 150 years.

In 2012, Pelkonen, Abass and Wiesner reaffirmed the toxicity of thujone—especially in high concentrations—but this does not prove the phenomenon of absinthism, in their view (*Regulatory Toxicology and Pharmacology* 2013, n.pag.). Interestingly, the chemical thujone, its effects on humans and the phenomenon of absinthism are still analysed without one definitive answer. Despite mass spectrometry and gas chromatography, it has not been possible to determine the actual effects of thujone on the body. What can be said, however, is that the concentration of thujone in historic absinthes was too low to show any toxic effect (Lachenmeier 2007, 308) and that absinthism was in fact alcoholism or alcohol poisoning (308). The current levels of thujone allowed in the EU are actually higher than anything Lachenmeier found (2006, 372).

It can be concluded from these differing findings that the myths and legends still influence the perceptions of absinthe today. To demystify them with scientific results is an ongoing process. Non-conclusive findings of contemporary research also help to illuminate nineteenth-century literary texts that deal with absinthe. In the following section, two texts are analysed in order to demonstrate how contradictory attitudes towards the beverage were in the nineteenth century. Christina Rossetti and Robert Hichens explore absinthe in different ways and reflect diverse and indeterminate understandings of absinthe's known drug effects. Rossetti's poem "Goblin Market" (1862) illustrates the dangers of addiction and the way to overcome it. However, the poem is not clear in its reading of potential dangers, as she uses metaphorical and ambiguous language. Hichens' novel *The Green Carnation* (1894), on the other hand, portrays absinthe as an element of Victorian society loved by some (Decadents, amongst others) and despised by others (conservatives), thus demonstrating the ambivalent perception of the beverage.

Christina Rossetti's "Goblin Market" (1862)

"Goblin Market" provides the reader with a dense imagery which is both poetic and leaves room for interpretation. Apart from the reading applied here, the poem has also elicited many interpretations from theological readings (Arseneau 1993) to the sexualised female body (Carpenter Wilson 1996). Deborah Ann Thompson reads the poem in terms of anorexia (1991), whereas Marylu Hill stresses the eroticism of devouring fruit (2005). Other interpretations include essays on different aspects of economics (Holt 1990), vampirism (Morrill 1990) or consumerism (Helsinger 1991). The poem revolves around two sisters, Laura and Lizzie, who encounter evil goblins. These goblins sell Laura some fruit, which makes her ill after tasting them. In order to save her sister from death, Lizzie faces the goblins trying to obtain a cure, but she is attacked. The goblins smear her face and body with fruit to make her eat them as well, but Lizzie prevails. Her resistance eventually wears her attackers down. On returning home, Laura kisses her sister's face and is miraculously cured by the fruit juices.

In this section, I suggest that the poem can also be read in terms of addiction, in particular in the context of absinthe consumption. Furthermore, it is the poem's open nature that contributes to the discussion of this chapter: "Goblin Market" and absinthe share an ambiguous discourse. In this context, one particular line is of interest: towards the end of the poem, Lizzie tries to save Laura from her demise and implores her to "Hug me, kiss me, suck my juices / Squeez'd from goblin fruits for you" (Rossetti 1893, 56). Laura obeys, but the sensation is different from her first encounter with the fruits: "Her lips began to scorch, / That juice was wormwood to her tongue, She loath'd the feast" (58). As Shelley O'Reilly argues, "Christina Rossetti, who wrote much about the language of flowers and other plants, would have been cognizant of the properties and effects of wormwood" (1996, 112). Rossetti uses the plant wormwood here as a metaphor for something that not only tastes horrible but also can be read as a poison.

As pointed out, wormwood had been used as a medicine well before the 1880s. In giving Laura an antidote for the addiction to the goblins' fruits, Lizzie cures her with a 'bitter' medicine. In other words, "[the] goblin fruit has become toxic to her senses but this has the effect of curing her because it is an antivenin: she is cured by a hair of the dog that bit her" (O'Reilly 1996, 111). In this context, wormwood can also be read as an addictive poison, which almost kills Laura. Although it tasted sweet and incredibly good at the beginning, it turned bitter after a period of abstinence (i.e. she is no longer physically dependant), and so she was cured. If read along these lines, the addictive substance works similarly to alcohol. As stated above, both Brown and Magnan believed that there were specific kinds of alcohol that can cure a patient. In Laura's case, wormwood is used here to stand in for a type of alcohol with a certain known quality.

One could argue that the fruits offered by the goblins at the poem's beginning represent not only sensual sins but also the temptations of alcohol and the dangers of addiction. As a tale-within-the-tale, the fate of another victim to the goblins, Jeanie, is used as an example: she tasted the goblins' fruit and then died from succumbing to this addiction. As has been pointed out by Foxcroft, chronic opiate use was common in nineteenth-century England (2007, 1). Rossetti's poem can thus also be read along the lines of opium addiction. Richard J. Miller and Phuong B. Tran interpret the poem accordingly: "Laura succumbs to temptation and eating the fruit produces an ecstatic drug-like effect. However, subsequent attempted encounters with the goblins or their fruits are ineffective (tolerance?) and are associated with the development of physical symptoms typical of opium addiction" (2000, 301). In addition to this reading, the subtle reference to addiction can also be linked to that of absinthe. The symptoms of addiction as described in Castelow's account of *Opium in Victorian Britain* (2017) were thought to include aches, cramps, nausea, vomiting and diarrhoea alongside deep depression, restlessness and slurred speech (*historic-uk.com*). Similar to alcohol addiction withdrawal, the symptoms seemed to affect the body and the mind alike. Rossetti's poem does not exactly state what kind of effects of consumption is meant, hence its persisting appeal for scholars interested in a variety of substances.

Along these lines, one can argue that Laura suffers from addiction after her return from the market. She has no other focus than obtaining the next 'taste' and is utterly disappointed when the goblins will not come for her anymore. This behaviour is consistent with addiction. Her longing is described accordingly:

> They sat up passionate yearning,
> And gnash'd her teeth for baulk'd desire,
> and wept
> As if her heart would break. (32)

The exceeding pain she experiences comes from the physical longing for more 'fruit', or in other terms, the symptoms of withdrawal from her chosen drug. However, Laura does not hear the goblins' call ever again. Miller and Tran associate this with developing a tolerance for the drug, a reading which, in the context of addiction to both opiates and alcohol, is not only plausible, but also likely.

The wormwood/absinthe reference is decisive, as it not only represents forbidden desire but also 'dangerous' habits. O'Reilly argues that Laura undergoes a kind of exorcism (111) when Lizzie makes her taste the fruits again. She writes: "Laura undergoes a purging, an exorcism in which her body convulses, her limbs writhe, her hair streams. Her body is wracked by a parasite ingested with the goblin fruit, and this metaphysical parasite, the parasite of desire for the fruit, is expelled with Laura's seizure – poison effects the cure" (111). This cure comes at the price of more physical pain, but the cure is permanent. The experienced pain can be read as the symptoms of withdrawal. After having gone through that ordeal, Laura remains 'clean'—from alcohol, absinthe or both—for the rest of her life.

O'Reilly actually links all of Laura's symptoms to absinthism: hallucinations, brain damage, sleeplessness, tremors and convulsions (110). The results of Hein's research would actually support this reading of both poem and illness. However, when considering the latest findings of Lachenmeier, it has to be noted that the symptoms also describe alcoholism—a fact that O'Reilly concedes, but she still maintains the existence of absinthism as the source for Laura's condition. Lachenmeier concludes in his article from 2007: "Recent research of [nineteenth-]century studies about absinthism concluded that the so-called absinthism cannot exactly be distinguished as a distinct syndrome from chronic alcoholism. [...] Nowadays it appears most likely that in the same way alcoholism was first called absinthism, because absinthe was one of the main alcoholic beverages consumed in that era" (Padosch, Lachenmeier et al. quoted in Lachenmeier 2007, 260). Thus, O'Reilly's argument may be seen as valid when it comes to Laura's alcohol addiction, but the illness may not have been called absinthism after all. It has to be said here that—as there is still an ongoing dispute about whether something like absinthism actually exists—one should be very vigilant with these concepts. As Lachenmeier points out repeatedly in his articles, the symptoms of both illnesses are similar; what differs is the active ingredient, namely thujone. As we have seen, the impact of thujone is equally disputed.

Robert Hichens' *The Green Carnation* (1894)

Hichens' short novel *The Green Carnation* shows the same ambivalent attitudes to absinthe and alcohol in general. The text provides a glance at the Decadent way of life and establishes absinthe-drinking as one of the circle's elements, amongst other activities (e.g. indulging in food, parties and excesses). Hichens' novel thus fits into the group of Decadent books because it celebrates and simultaneously represents as scandalous the lifestyle of its followers, and it gives a vivacious account of how Oscar Wilde's circle operated, depicting the life of Wilde and his relationship with Lord Alfred Douglas. The first publication in 1894 was anonymous and many readers speculated about Wilde being the author himself.

This work of biographical fiction paints a picture of bohemian London and how it was perceived from the outside. Hichens, who was part of Wilde's circle of friends, "may have wished to ingratiate himself to Wilde by portraying him so cleverly" (Beckson 2000, 391). Although the text does not include many direct references to absinthe, the few paragraphs that mention the drink are all the more telling. It is Reggie Hastings, the disguised Lord Alfred Douglas, who confesses first to drinking absinthe. In a conversation with Lady Locke and Mrs. Windsor about sensations of any kind, he states that at times he likes to be prudent and prefers simple enjoyments, such as reading or almost ascetic meals. But he also enjoys the other side of the spectrum as the following paragraph shows:

> At other times I must drink absinthe, and hang the night hours with scarlet embroideries. I must have music, and the sins that march to music. There are moments when I desire squalor, sinister, mean surroundings, dreariness, and misery. The great unwashed mood is upon me. Then I go out from luxury. The mind has its West End and its Whitechapel. The thoughts sit in the Park sometimes, but sometimes they go slumming. They enter narrow courts and rookeries. They rest in unimaginable dens seeking contrast, and they like the ruffians whom they meet there, and they hate the notion of policemen keeping order. (Hichens 1894, 5)

In this passage, Hichens offers a description of the effects of absinthe drinking, which is associated with social decline and crime. The more one consumes, the more attractive the 'darker' sides of society become. Although Reggie only speaks of his thoughts, he makes clear that his "mind govern[s] the body" (*Green Carnation*, 5), hence, not only his mind becomes debased, but also his body. Apart from the reference to absinthe, another popular drug of the Victorians can be found here: the "unimaginable dens" may be read as a reference to places to consume opium in more disreputable parts of London. Both addictions are presented together as a picture of social digression. For the temperance movement, the social standing of opium addicts was no different from that of alcohol addicts, and this view of addiction was one of moral and mental weakness leading automatically to social decline.

Such was the context in which absinthe came to England from France where it had been quite popular already. The English bohemia in London used the drink to set themselves apart from the gin-drinking population and as a way to embrace French culture, which they aimed to copy. Therefore, artists might have embraced absinthe as their token drink, but for the rest of the population it "was a visible, poisonous symbol of this French encroachment on British art, morality and territory" (Adams 2004, 8). One could say that in this case the concept of addiction was not only about race and class, but even more so about nationality. Hence, the Victorian establishment, which was shocked by the Decadence movement already, found a scapegoat in the form of absinthe; they could pin any sin or misbehaviour on the French drink. Although the drink was immensely popular in France, it never reached these levels of popularity in Britain. Consequently, there was never a need for a ban. However, in the long run, the strong sentiments against absinthe might have affected the views on the drink's drug effects.

The social, cultural and national contexts are also of importance when discussing the occurrence of the beverage in *The Green Carnation*. Absinthe also plays a role for the Wilde-like character Esmé's lifestyle. As an explanation for the carnation's colour, Mrs Windsor states: "He wore it, in the first instance, because it blended so well with the colour of absinthe" (Hichens 1894, 10). The direct reference to the signature drink of the Decadents put the emblematic flower at the centre of the discussion, alluding to another now problematic statement about social decline. According to Karl Beckson, "in the novel, the flower is associated with 'unnatural sin' (never defined by Hichens but sufficiently obvious, one presumes, to most reviewers and to informed readers)" (2000, 392). The carnation as such is an 'unnatural' creation which combines two plants; hence, one can interpret the carnation as an allusion to conservative rhetoric against homosexuality in the Victorian period, weaving together two different conservative concerns about social decay.

The most telling paragraph on Esmé's connection to the drink is part of his rant about intelligence:

> [I]ntelligence is the demon of our age. Mine bores me horribly. I am always trying to find a remedy for it. I have experimented with absinthe, but gained no result. I have read the collected works of Walter Besant. They are said to sap the mental powers. They did not sap wine. Opium has proved useless, and green tea cigarettes leave me positively brilliant. What am I to do? I so long for the lethargy, the sweet peace of stupidity. (Hichens 1894, 47)

Although this oration is to be taken as an ironic outburst, he draws on the widespread notion that drinking absinthe 'degenerates' one's mind and leads to lethargy. By using irony, Esmé ridicules the prevalent point of view of conservatives towards the more open or artistic lifestyles (like that of the Decadents). The notions mentioned in his speech reflect Magnan's medical studies, as well. It was known that the French feared a 'degeneration' of their population due to different types of alcohol. Again, lethargy and stupidity are associated with absinthe.

In the light of twenty-first-century research on absinthe, Hichens' novel not only exemplifies the lifestyle that surrounded this particular drink in late Victorian England but also highlights the contradictory medical opinions and myths about it. Differing perspectives on absinthe are reflected by Reggie/Esmé and Lady Locke in the novel. Esmé states that absinthe dulls the mind, and that is one effect frequently attributed to it. Esmé's remark implies that consumption leads to stupidity and thus he adheres to Magnan's reading of the drinkers' symptoms. But in the light of Lachenmeier's findings, Esmé is wrong: it is not the absinthe or the thujone, but rather the alcohol itself doing the damage. Reggie succumbs to the appeal of the legend and uses it to complete his nightly ramblings as part of the bohemian circles he is a part of, and he embraces the drink with all its myths in tact. He does not seem to be critical about it at all.

All in all, the novel provides a highly ironic insight into the life of the Decadents surrounding Wilde in the late Victorian era. As a close acquaintance of Wilde and Douglas, Hichens certainly knew many anecdotes about them, but also their attitudes towards certain topics (the consumption of absinthe being one of them). For the evaluation of absinthe, the novel shows that the myths surrounding absinthe were part of its consumption and that the drinkers knew the legends of its effects. In other words, Hichens was aware of the cultural and social baggage the consumption of absinthe came with, and *The Green Carnation* celebrates the Decadents rather than supporting the discourse on 'degeneration'. The indeterminate drug effects allow for a multitude of assumptions and legends to be associated with absinthe. It is this particular indeterminacy that is reflected in the literary texts of the late Victorian era. Both Rossetti's and Hichens' texts serve as valuable contributions to the discourse on absinthe/absinthism as well as addiction, as they reveal how differently these issues were treated.

Cultural Aspects of Absinthe

Pharmacological analyses on absinthe focus mainly on the one element that seems to be responsible for distinctive effects: thujone. It has become the focal point of chemical studies, but before this research, the substance was also the basis for many persistent legends: so persistent, in fact, that scientists accuse each other of not having moved on from the notions of the nineteenth century, as Lachenmeier's response to Hein shows. Thujone has a similar molecular structure a s cannabis, but does not show all the same effects. It is highly toxic in concentrated form, as the experiments on animals have shown. But, when thujone is part of absinthe, it is diluted and therefore not toxic anymore. The illness of absinthism may have been a form of alcoholism and thus due to excessive alcohol consumption.

What is important in the case of absinthe is also the cultural aspect. The circulation of legends contributes much to the reputation of absinthe as a supposedly 'dangerous' drink. Therefore, the discourse about absinthe should be seen in context; reading literature about absinthe helps us to see the different ways in which drug effects could and can be interpreted. The analysis of nineteenth-century literature and culture alongside twenty-first-century scientific studies has shown that there was not and is not yet a definite answer to the drug effects of thujone in absinthe. It can be said, however, that the belief in absinthe's effects significantly contributes to the legend and thus the interpretation of it, working as a kind of placebo effect. And it is probably safe to say that the myths of the *fée verte* will remain a part of popular culture, as many today still believe in absinthe's distinct power.

References

Adams, Jad. 2004. *Hideous Absinthe: A History of the Devil in a Bottle*. London and New York: I.B. Tauris.

Arseneau, Mary. 1993. Incarnation and Interpretation: Christina Rossetti, the Oxford Movement, and 'Goblin Market'. *Victorian Poetry* 1: 79–93.

Beckson, Karl. 2000. Oscar Wilde and the Green Carnation. *English Literature in Transition, 1880–1920* 43 (4): 387–97.

Brodie Farrell, Janet, and Marc Redfield. 2002. Introduction. In *High Anxieties: Cultural Studies in Addiction*, ed. Janet Brodie Farrell and Marc Redfield, 1–15. Berkeley, Los Angeles and London: University of California Press.

Carpenter Wilson, Mary. 1996. 'Eat Me, Drink Me, Love Me': The Consumable Female Body in Christina Rossetti's Goblin Market. In *Victorian Women Poets*, ed. Tess Cosslett, 212–33. London: Routledge.

Castelow, Ellen. n.d. Opium in Victorian Britain. *Historic-uk.com*. http://www.historic-uk.com/HistoryUK/HistoryofBritain/Opium-in-Victorian-Britain. Accessed 10 August 2017.

Foxcroft, Louise. 2007. *Making of Addiction: The 'Use and Abuse' of Opium in Nineteenth-Century Britain*. Hampshire: Ashgate.

Hein, Jakob, Lars Lobbedey, and Klaus-Jürgen Neumärker. 2001. Absinth – Neue Mode, alte Probleme. *Deutsches Ärzteblatt* 98 (42): 2716–24.

Helsinger, Elizabeth K. 1991. Consumer Power and the Utopia of Desire: Christina Rossetti's 'Goblin Market'. *ELH* 58 (4): 903–33.

Hichens, Robert. 2015 [1894]. *The Green Carnation*. McAllister Editions.

Holt, Terrence. 1990. 'Men Not Such in Any Town': Exchange in 'Goblin Market'. *Victorian Poetry* 28 (1): 51–67.

Judžentienė, Asta. 2016. Wormwood (*Artemisia absinthium* L.) Oils. In *Essential Oils in Food Preservation, Flavour and Safety*, ed. Victor R. Preedy, 849–56. Amsterdam: Elsevier.

Lachenmeier, Dirk W. 2007. Absinthe—History of Dependence to Thujone or Alcohol? *Fortschritte Neurologie Psychiatrie* 75: 306–08.

Lachenmeier, Dirk W., Stephan G. Walch, Stephan A. Padosch, and Lars U. Kröner. 2006. Absinthe—A Review. *Critical Reviews in Food Science and Nutrition* 46: 365–77.

Lachenmeier, Dirk W., Nathan-Maister, David, Theodore A. Breaux, Jean-Pierre Luaute, and Joachim Emmert. 2010. Absinthe, Absinthism and Thujone—New Insights into the Spirit's Impact on Public Health. *The Open Addiction Journal* 3: 32–38.

Magnan, Valentin, and A. Fillassier. 1912. Alcoholism and Degeneracy. *Problems in Eugenics: Papers communicated to the First International Eugenic Congress*. London: The Eugenics Education Society.

Miller, Richard J., and Phuong B. Tran. 2000. More Mysteries of Opium Reveal'd: 300 Years of Opiates. *Trends in Pharmacological Sciences* 21 (8): 299–304.

Morrill, David F. 1990. 'Twilight Is Not Good for Maidens': Uncle Polidori and the Psychodynamics of Vampirism in 'Goblin Market'. *Victorian Poetry* 28 (1): 1–16.

O'Reilly, Shelley. 1996. Absinthe Makes the Tart Grow Fonder: A Note on 'Wormwood' in Christina Rossetti's 'Goblin Market'. *Victorian Poetry* 43(1) (Spring): 108–144.

Patočka, Jiří, and Bohumil Plucar. 2003. Pharmacology and Toxicology of Absinthe. *Journal of Applied Biomedicine* 1: 199–205.

Pelkonen, Olavi, Khaled Abass, and Jacqueline Wiesner. 2013. Thujone and Thujone-Containing Herbal Medicinal and Botanical Products: Toxicological Assessment. *Regulatory Toxicology and Pharmacology* 65 (1): 100–07.

Prestwich, Patricia. 1979. Temperance in France: The Curious Case of Absinth. *Historical Reflections/Réflexions Historique* 6 (2): 301–19.

———. 1994. Drinkers, Drunkards, and Degenerates: The Alcoholic Population of a Parisian Asylum, 1867–1914. *Histoire Sociale/Social History* 27 (54): 321–35.

Risse, Guenter B. 1988. Brunonian Therapeutics: New Wine in Old Bottles? *Medical History* 8: 46–62.

Rossetti, Christina. 1893. *Goblin Market*. London: Macmillan. http://digitalcollections.baylor.edu/cdm/ref/collection/ab-wpc/id/85127. Accessed 22 January 2020.

Semmler, F.W. 1894. Ueber Tanaceton und seine Beziehungen zu Thujon. *Berichte der deutschen chemischen Gesellschaft* 27: 895–98.

Trotter, Thomas. 1988 [1810]. *An Essay, Medical, Philosophical, and Chemical, on Drunkenness and Its Effects on the Human Body*. London: Longman.

Vickers, Neil. 1997. Coleridge, Thomas Beddoes and Brunonian Medicine. *European Romantic Review* 8 (1): 47–94.

Youngquist, Paul. 1999. Lyrical Bodies: Wordworth's Physiological Aesthetics. *European Romantic Review* 10 (1-4): 152–62.

Zieger, Susan. 2008. *Inventing the Addict: Drugs, Race and Sexuality in Nineteenth-Century British and American Literature*. Amherst: University of Massachusetts Press.

PART IV

Historicising the Prescription: Medication and Self-Medication

CHAPTER 11

"She Furnishes the Fan and the Lavender Water": Nervous Distress, Female Healers and Jane Austen's Herbal Medicine

Rebecca Spear

In a process of revision lasting from 1792 to around 1816, someone purloined Kitty's lavender water. The heroine of Jane Austen's novella "Kitty, or the Bower" uses and carries the preparation to a ball. Yet, at some point in the text's history, the allusion was deleted (Sutherland 2010; Austen 1792, 70). Its inclusion presents an image of a teenage girl, a fictional contemporary of Austen, who wields a plant-based remedy that symbolises women's control over manifestations of emotional distress as well as their agency as healers in the home. Its deletion suggests that the heroine should instead become solely occupied with perfecting an acceptable, attractive appearance prior to her entrance into a social and sexual marketplace. Kathryn Sutherland asserts that the manuscript of *Volume the Third*, of which "Kitty" forms the largest part, contains several hands (Sutherland 2015, 124). Positing

R. Spear (✉)
Cardiff University, Cardiff, Wales, UK
e-mail: spearrs1@cardiff.ac.uk

N. Roxburgh and J. S. Henke (eds.), *Psychopharmacology in British Literature and Culture, 1780–1900*, Palgrave Studies in Literature, Science and Medicine,
https://doi.org/10.1007/978-3-030-53598-8_11

connections between the volume and evidence of Austen's support for the literary endeavours of her niece and nephew, Anna Lefroy and James Edward Austen-Leigh, Sutherland suggests that the manuscript was a space of familial collaboration—a record of early experiences of composition and editing shared by aunt, niece and nephew (Sutherland 2015, 126). Although it is difficult to identify conclusively the hand that deleted the allusion to lavender water—it could have been James Edward or Jane Austen—most alterations appear to have been made by Austen's nephew during his adolescence in 1815 or 1816 (Sutherland and Johnston 2017a, xxii–xxiii; Sutherland and Johnston 2017b, 240). Such an editorial decision, however, becomes compelling because of its resonance with coexisting narratives of domestic medicine. Here, the "improving" hand of a younger generation (Sutherland 2015, 141) altered a manuscript, removing a potentially outdated allusion to a young woman using a herbal remedy. The deletion recreated the marginalisation of the female healer and associated lay practices by "medicalization" at the hands of male professional practitioners (Porter 1992, 101; Schiebinger 1991, 116). This erasure of female healers and denigration of their knowledge of herbal medicine often manifested in the removal of items from the *materia medica* and cautions over use of associated remedies (Cullen 1789, 1:143). The manuscript thereby forms connections between fiction and medicine, biography and historiography. Sutherland and Johnston's edition of the juvenilia restores the allusion to "Lavender water" (Austen 2017, 184). In so doing, it allows Jane Austen's original meaning to stand on the printed page, giving the reader a glimpse into real and imagined spaces where herbal medicine denotes proto-feminist and communal enfranchisement.[1] This essay will examine Austen's awareness of the experimental *materia medica* and the efficacy of herbal medicines recommended for nervous symptoms. Focussing on lavender, it will contend that "Kitty, or the Bower", *Northanger Abbey* and *Sense and Sensibility* feature allusions to its preparations as Austen interrogates prescriptive, fashionable gender identities and endorses the practice of rational female healers who use herbal remedies, assess their effects and alleviate the acute nervous distress of their friends and sisters.

[1] Doody's edition *Catharine and Other Writings* omits lavender water (2009, 204), as does Bree, Sabor and Todd's *Jane Austen's Manuscript Works* (2013, 182). R.W. Chapman's editions, *Volume the Third* (1951, 73) and *The Works of Jane Austen* (1963, 213), place lavender water in brackets.

The Pressures of Fashion: Medicinal Meets Scientific Botany in "Kitty, or the Bower"

The eighteenth century saw an evolution in women's engagement with science. Instead of practising medicinal botany, women were increasingly encouraged to undertake the study of scientific botany, with its emphasis on the controlled tasks of "collecting, naming and contextualising" (Dolan 2008, 120). Such alterations to the framework of European polite female education partially effected the estrangement of women from the remedies and practices that had contributed to their roles as lay practitioners, restricting what had been regarded as a necessitous understanding of the medicinal properties of plants, to engagement with a scientific discipline conveniently classified as a pastime suitable for women.[2] In "Kitty, or the Bower", Austen registers this shift between botanical knowledges, marrying or synthesising both disciplines and asserting the continued relevance of medicinal botany, or herbal medicine, as a lay practice which sustains female communities and promotes efficacious medical care. Herbal medicine remains central to women's experience of healthcare, as patients and practitioners, challenging and refuting contemporary marginalisation of the female healer.

Austen interrogates this movement from medicinal to scientific botany in "Kitty" by analysing the ways in which appreciation of an enclosed garden space, usually associated with the scientific study of botany, can provide agency and emotional support for young women through its associations with herbal medicine. A collective project undertaken by Kitty and her friends, Cecilia and Mary Wynne (2017, 169), the Bower provides refuge from the social and sexual expectations of others. Kitty's preoccupation with the Bower aligns her with the naturalist aware of the classification of plants, but also with female healers practising herbal medicine. The "constant releif [sic]" that Kitty finds in the Bower (169) creates an image of a young woman successfully self-healing through access to a plant-filled space that replicates the effects of remedies designed to ease nervous symptoms. It extends perceived

[2] See Shteir (1996, 38–39), Cohen (2006, 321; 331), Lynch (2010) and Stone (1990, 232–33). These pedagogical, cultural and social shifts are well documented. Londa Schiebinger (1991, 112–18) and Valérie Lastinger (2014, 130–33) have discussed the effects which contemporary historiographies of science, as well as manuals and pamphlets written predominantly by men, had upon the marginalisation of female healers.

healing properties that "tranquilize" Kitty's "mind and quiet her spirits" (170), easing her loneliness and distress following the Wynnes' departure, and thus soothes through its associations with female friendship and individual enfranchisement (Page and Smith 2011, 215-19). The Wynnes' connection to the Bower resurrects and vindicates a past where female community underpinned health care and, by extension, happiness. Austen's allusions to the "charm" (169, 182) which the Bower and the sisters wield over Kitty's "spirits" (170) recall folk remedies, forging a symbolic relationship between the Bower and restoration of the female healer's reputation and knowledge. Consequently, Kitty's insistence that "her Bower alone could restore her to herself" (170) creates associations between its "[s]hrubs [...] rendered sacred" (171) and medicinal properties, female friendship and individual agency.

Austen's writings were composed and revised while she was living in collaborative communities (Sutherland 2015, 126; 141) focussed upon shared experiences of healing underpinned by editing and circulation of manuscripts literary and medical. Steven King has uncovered interconnected medical communities comprised of professional and lay practitioners disseminating knowledge and dispensing medicines to patients (2006, 61). Steventon and Chawton in Hampshire (Tomalin 1998, 87; 215–17), as well as Godmersham in Kent and London, comprised similar, expansive neighbourhoods that surrounded Austen and circulated information about health in letters and conversation (Tomalin 1998, 78). Austen's knowledge of herbal remedies used to treat nervous complaints would have originated from household books, advice manuals, pharmacopoeias and the community.

In Austen's lifetime, William Cullen's nosology was challenged by his former student, John Brown, whose theory of disease dismantled and simplified Cullen's system (Budge 2013, 12–13; 56–57). Brunonian medicine connected the impact of stimuli upon the body's processes and health (Brown 1795, 1:4–5; 21), somatising nervous conditions and dividing disease into two classes, asthenic and sthenic (1:52–53). Characterised by debility (Beddoes 1795, 1:xxi), asthenia was caused by exposure to stimuli that resulted in "deficiency, of exciting power" (1:xxvi–xxvii). Sthenia originated in its accumulation or "excess" (1:xxvi–xxvii) due to poor stimulation or inactivity. The "powers" that caused the development of either class "were the remedies of the other" (1:xxvi–xxvii) as health depended upon equilibrium. Consequently, a debilitating, "weaker stimulus" (Brown 1795, 1:79) improved sthenic diseases and greater "stimulants" (Beddoes 1795, 1:xxv) relieved asthenic diseases (Brown 1795, 1:80). Like Brown, Erasmus Darwin purported that the "sensorium" or nervous system connected to other organs and was responsible for health (2009, 1:II.ii.i 10). The body's "sensorial powers" (1:XII.ii.i 75) worked in unison or isolation, beginning with irritation by a stimulus (1:II.ii.xiii 61) that provoked feeling remembered through association (1:VI.i–ii 34–36). This response could become an "acquired habit" (1:XII.iii.i 83) that might improve health (1:XII.vii.iii 96–97) or produce "intoxication" (1:XII.iii.i 82) if a stimulus was repeated too often and in large doses (1:XII.ii.i–iii.i 80–82). Such theories about the effects of stimulants upon the body, their production of health, debility and habit, corresponded with early ideas concerning addiction that transformed between the seventeenth and nineteenth centuries, evolving from a language of substance appreciation to abuse (Zieger 2008, 3–6; 23).

Women, including Austen, mention lavender water in their letters and diaries, emphasising its importance in the machinery of everyday life. Austen's neighbour, Eliza Chute, declared in the "Accounts of Cash" section of her 1797 diary that "To lavender water" went "5 Shillings" (n.pag.). In 1801, Jane Austen asked her sister, Cassandra, to "purchase for [Martha Lloyd] two bottles of Steele's Lavender Water when you are in town" (Austen 2011, 75). Although the Austen sisters and their close friend, Martha, might have made lavender water at home, these purchases suggest convenience as well as the effects of fashion and "medical consumerism" (Hill Curth 2006, 29; 37–39). While "home-made medicines would have been more economical [than proprietary medicines or personalised prescriptions]" (41), in the midst of a growing, competitive medical marketplace that encouraged purchase over manufacture, "kitchen physick [sic]" saw "a gradual erosion" in the face of "demand for proprietary drugs" (Hill Curth 2006, 46; see Schiebinger 1991, 116; Vickery 1999, 157; Stobart 2016, 120–21). *Materia medicas* gave botanical, medicinal and pharmaceutical information about lavender, explaining that it could be bought in proprietary medicines, dried and found in kitchen gardens (Boerhaave 1755, 158–59).

Lavender was used widely as a cosmetic (Black and Le Faye 2002, 36) and popular remedy in the treatment of nervous diseases (Beatty 2012, 126). Known as a simple medicine, it was "composed of or concocted from one constituent" (Stobart 2016, 196). William Cullen's *Treatise of the Materia Medica* (1789) stated that lavender was "of this order named Cephalics" (2:147) suitable for the treatment of complaints affecting the brain and nerves because it stimulated the nervous system (2:148). Accepted as an efficacious folk remedy (Wiltshire 2005, 310), the properties and uses of lavender were published by and for the medical faculty and doctored for a public increasingly preoccupied with their health under the directions of professional practitioners. Lavender's availability, properties and popularity reflect its status as a fashionable preparation and proprietary medicine used by those with acute nervous symptoms.

In the eighteenth century, self-care encouraged individuals to manage their own health. Preventative and curative measures meant that an enthusiastic public wielded medical knowledge as a means to maintain and restore health (Porter 1992) and, coinciding with a cultural and medical focus upon nervous disease, remedies were inevitably sought. Published advice ensured that knowledge of "rational" medicine (Porter 1992, 106; 108) allowed the public greater access to the aetiology and epidemiology of their disease experience, creating "deference and dependence on specialised knowledge" (Smith 1985, 272). Published manuals invited the reader's engagement with physiology and medicine while controlling domestic healing through prescriptive regimen (Porter 1992, 110–11; King 2006, 62; 68). At this time, irregular practitioners, including women, became subject to regular "male professional opposition" (Smith 1985, 273). Medical men often addressed their works to women in the hope that, re-educated, they would leave behind their supposedly ignorant reliance on folk knowledge (Buchan 2014, 8; 45; xiii).

Austen's "Kitty" transcribes such conversations between culture and medicine, with its allusion to lavender water showing an awareness of the ways in which women were encouraged to display sensibility and to treat its nervous symptoms by self-dosing with remedies professionally sanctioned. Before her departure for a ball, hurried adjustments see Kitty's maid, Nanny, "dispatched in the same haste to order the Carriage, while her Mistress was putting on her gloves, arranging the folds of her dress, and providing herself with Lavender water" (Austen 2017, 184).

For much of the eighteenth century, smelling-bottles contained vinegars, salts comprised of ammonia or hartshorn and preparations like lavender or Hungary water (Friedman 2016, 52; Cullen 1789, 2:151). Initially, "personal scenting devices" (Friedman 2016, 59) were carried to mask "bad air" considered an affront to society and detrimental to health (53–54). Yet, as the century progressed, smelling-bottles became objects connected with nervous disease. Regarded partly as a "fashionable accessory" (Friedman 2016, 56) carried by those wishing to display sensibility but also to alleviate nervous symptoms (54), the smelling-bottle was a symbol of the performativity of emotional refinement (Friedman 2016, 55, 66) and the preserve of those who experienced the uncertain reality of nervous disease (Beatty 2012, 66). Consequently, such items were regarded as "for emergency use only" (Friedman 2016, 65–66), a necessity to be "identified either by the user or by immediate observers" (54) in moments of acute nervous distress "remedied by reviving smells" (Friedman 2016, 55). Emily Friedman contends that Austen often depicts "salts and smelling-bottles" (66) as "ineffective" (66), making the devices and characters who use them "objects" of "satire" (67). However, Austen's portrayals of fashionable remedies are complicated by evolutions in material culture and medicine. Victims of hypothetical sexual difference, her fashionable ladies and rational women self-dose, dispense or recommend unthreatening, appropriate remedies, like lavender, that dismantle associated gender identities by acting upon the female body and recovering its health and strength. Lavender's soothing, revivifying effects upon the user serve as polemic against the intellectual and emotional subjugation of female patients and practitioners.

Kitty's lavender water (in its smelling-bottle) will mask stenches, protect from disease and soothe nervous distress as a fashionable, "emergency remedy" (Friedman 2016, 56). In this way, the novella is influenced by contemporary discourses that connected sensibility, medical prophylaxis and nervous disease (Porter 1987, 182). By wielding lavender water to self-dose and oversee others' care, Kitty becomes synonymous with the continued necessity of herbal remedies for nervous distress, symbolising women's protests against sexual difference and its effects upon their marginalisation. Mirroring the Bower's properties, Kitty's lavender water has the potential to upset or subvert medicalisation as a folk substance (Raffald 1788, 368) whose efficacy was accepted by the medical faculty (Cullen 1789, 2:147–48). Kitty's connection to garden spaces and herbal medicine functions as a reminder of the restorative role of plants in the alleviation of nervous symptoms. Her dependence upon emotional and botanical knowledges promotes the survival of folk medicine, or kitchen physic. Harnessing the potency of plants to protect and heal herself, Kitty vindicates women's medical knowledge and practice and recovers proto-feminist female communities, the novella thematically foreshadowing its own restoration. These intersections between fashionable ideals and women's management of medical care offer a preface to Austen's portrayals of female healers, their oppression and enfranchisement.

Healing at Home: Friendship, Matriarchy and the *Materia Medica* in *Northanger Abbey* (1818)

In *Northanger Abbey*, lavender becomes a metonym for women's marginalisation from contemporary medical knowledge and practice. A dialogue between the novel's protagonists, Catherine Morland and Henry Tilney, about dance partnerships and marriages (Austen 2008a, 54) advances Austen's examination of the effects of fashion upon the practice of herbal medicine in "Kitty".[3] Henry asserts that: "In marriage, the man is supposed to provide for the support of the woman; the woman to make the home agreeable to the man; he is to purvey, and she is to smile. But in dancing, their duties are exactly changed; the agreeableness, the compliance are expected from him, while she furnishes the fan and the lavender water" (55). For Henry, lavender water is a preparation that denotes feeling and display of emotion, but also enables the demarcation of separate spheres with their associated gender identities. These prescriptive roles or attributes were underpinned by theories of sexual difference, which often centred upon the nerves (Beatty 2012, 142; 158–60; Barker-Benfield 1992, 247). In the eighteenth century, discourses of sensibility argued that the nerves were vital to the heightened emotional responses noted as markers of superiority and refinement (Porter 1987, 182). Connected with ideals of femininity, physiology and anatomy subjected women to subordination under nerve doctors and quacks (Beatty 2012, 41–61), who considered them physically susceptible to acute and chronic symptoms of nervous disease (Barker-Benfield 1992, 25–28; Dyckman 2014, 239). By Austen's birth in 1775, women's herbal knowledge had been marginalised and transformed into domestic skills (Schiebinger 1991, 116; Shteir 1996, 37–39). Austen interrogates attitudes to women's medical practice arising from such phenomena, where, equivalent to a decorative fan, lavender water had become a fashionable consolation for female healers.

[3] Clara Tuite has argued that *Northanger Abbey* is a later reworking of "Kitty" (2002, 27).

Consequently, albeit incidentally, Henry's comments maintain women's control of domestic medicine as Austen's image of a woman "furnish[ing]" the lavender water counters the divergence between female healers, herbal medicine and pharmacology. In this way, Austen's language echoes Elizabeth Blackwell's hope that her *Curious Herbal* (1737–39) would prove "useful to such as are not furnished with other[s]" (1:Introduction) as well as Mary Wollstonecraft's assertion that woman could only be dignified, reasoned and moral if her "head is furnished with ideas" (2008, 206). Allusions to such texts, connected with proto-feminist transmission of knowledge, preserve lavender water for an anticipated future of enfranchisement. Although medicinal botany was becoming increasingly medicalised and associated with professional practice, the dispensation of simple medicines for acute nervous distress remained synonymous with the medical care intrinsic to women's social interactions and private experiences.

Catherine's actions attest to the significance of proto-pharmacological knowledge, neural medicine and medicinal botany to women's agency as well as social and nervous health. Female "[f]riendship is certainly the finest balm for the pangs of disappointed love" (Austen 2008a, 20), providing a means of support and acting as a mechanism for medical care throughout the novel. Catherine's actions and discourse repeatedly remedy the nervous symptoms of her friends; she observes and attempts to quieten what she believes are Isabella Thorpe's genuine fears (89) and eases Eleanor Tilney's sense of isolation (132). This ability to remedy nervous distress culminates in dispensing lavender water to Eleanor. After hearing someone outside her chamber knocking, Catherine

> stepped quietly forward and opened the door. Eleanor, and only Eleanor, stood there. Catherine's spirits however were tranquillized but for an instant, for Eleanor's cheeks were pale, and her manner greatly agitated. Though evidently intending to come in, it seemed an effort to enter the room, and a still greater to speak when there. Catherine, supposing some uneasiness on Captain Tilney's account, could only express her concern by silent attention; obliged [Eleanor] to be seated, rubbed her temples with lavender-water, and hung over her with affectionate solicitude. (164)

Austen's description of a plant-based preparation applied to the temples follows advice on the treatment of "swoonings" in published medical manuals, administering a substance considered efficacious by folk and rational medicines, physicians and female healers (Buchan 2014, 554–55). The semicolon punctuates Catherine's thoughts and actions, its slight pause emphasising her concern and prompt response to Eleanor's "pale" complexion and agitation, both of which were nervous symptoms (Barker-Benfield 1992, 15; Porter 1987, 179; Cheyne 1733, 102). This readiness to diagnose, awareness of appropriate remedies used to strengthen the nerves and willingness to soothe Eleanor's distress indicate Catherine's understanding of medicine, acquired through reading and from experiences at Fullerton, Bath and Northanger (Austen 2008a, 8). Her application of a plant-based remedy to alleviate Eleanor's nervous distress compounds friendship with medicinal botany as integral to female communities, making her a symbol of Austen's protest against women's marginalisation within medicine. Thus, Austen's portrayals of Catherine as a healer and Eleanor as a patient converse with didactic depictions of rational, empathic female healers in contemporary fiction, such as Mary Wollstonecraft's novels (2009, 18).

Eleanor responds to Catherine's ministrations, declaring: "I am quite well. This kindness distracts me – I cannot bear it – I come to you on such an errand!" (Austen 2008a, 164). Austen's punctuation and language convey that Eleanor's distress is intensified to distraction, to derangement, rather than alleviated by Catherine's attention.[4] The lavender water's action upon Eleanor's brain and nerves (Cullen 1789, 2:147–48; Quincy 1719, 77–78) is impeded and its effect lessened because her father, General Tilney's, instructions regarding Catherine's expulsion from their home are unwarranted, humiliating and confusing. Eleanor considers herself undeserving of assistance or pity and, until the formation of "connected words" (Austen 2008a, 164) that disclose the General's demands and quieten her concern for Catherine, the lavender water's efficacy is delayed and does not act visibly upon the nerves by calming her expression or behaviour. Explaining the "errand", Eleanor's powerlessness and agitation become evident through displays of fear and resentment when discussing her father: "Her voice faltered, and her eyes were turned to the ground as she mentioned his name" (2008a, 164–65). Continually "endeavouring to collect herself and speak with firmness" (165), Eleanor's distress resurfaces and counters the lavender's actions, making

[4] Michelle Faubert has argued that dashes express mental anguish in Wollstonecraft's writings (2005, 139).

her inarticulate increasingly due to an uncomfortable, fearful acknowledgement of her father's autonomy. Catherine's assumption concerning Captain Tilney's involvement in his sister's distress, as well as General Tilney's culpability, implies that nervous distress is defined by men's actions and their deleterious effect on women's health and ability to heal.

General Tilney's effect upon the lavender's pharmaceutical efficacy replicates the expulsion, or reduced usage, of items from the *materia medica* by medical men. The findings of iatrochemical and iatromechanical experimentation (Maehle 1999, 4–8), in addition to the overt scepticism with which all medicine, especially herbal remedies, was approached (Loudon 1986, 62–65; Tullett 2019, 92–94), influenced the medical faculty's consideration of medicinal use and pharmaceutical inefficacy (Lewis 1784, v; Cullen 1789, 1:142–52) published in revised catalogues, such as the *Pharmacopoeia Londinensis* (Royal College of Physicians of London 1746) and *Pharmacopoeia Edinburgensis* (Royal College of Physicians of Edinburgh 1744). Jane Austen's brother, Edward Knight's, library at Godmersham Park included editions of these pharmacopoeias, providing evidence that the family had access to physicians' experimental, proto-pharmacological knowledge.[5] Londa Schiebinger contends that women's knowledge of herbal remedies was subject to bioprospecting akin to that taking place in the colonies (2004, 96–97). Drug testing contributed to the process of undermining women's knowledge of medicinal botany to medicalise and professionalise a male-dominated *materia medica*. Female healers increasingly encountered the intrusion of botanists and medical men who criticised and appropriated their remedies, leaving them uncredited (Schiebinger 2004, 96; 98). Austen's letters provide evidence of her awareness of folk remedies no longer sanctioned, detailing Mrs Austen's acquisition of a receipt for dandelion tea and its adaptation by local practitioner and family acquaintance, Mr John Lyford.[6] The receipt embraces lay traditions transmuted during this period of proto-pharmacological investigation (Francia and Stobart 2014, 12) that excluded popular, "useless substances" (Cullen 1789, 1:145) from pharmacopoeias (Aikin 1784; Worth Estes 1997, 37–44). It evidences that Austen lived amongst a community that continued to

[5] A digital reconstruction of the library, *Reading with Austen*, offers insights into the volumes with which Jane Austen would have been familiar (Sabor 2018).

[6] For more on Mr Lyford, see Austen (2011, 16), Boorman (1966, 36–40), and Tomalin (1998, 78).

use remedies appropriated or discredited and understood the importance of folk medicine to women.

In Austen's novel, patriarchal authority, symbolised by General Tilney, interferes with the Abbey's female community by controlling happiness and impeding psychopharmacological efficacy. The General's attempts to restrict Catherine and Eleanor's access to private spaces connected to his wife (Austen 2008a, 131–32; 136; 140–41) devalues female relationships. Revealed through her inarticulacy (164), Eleanor's visible distress betrays that terror exists in women's subjection to patriarchs who attempt to deny matrilineage and repress unregulated female friendship. Catherine and Eleanor are separated from reminders of the importance of matriarchy in a manner analogous to the ways in which the female healer's herbal knowledge and practice were undermined by medical men, who mocked their "pretended experience" and the "falsehoods" of "old women's tales" (Cullen 1789, 1:143). Yet, "the General's evident desire of preventing" access to spaces connected to his wife functions as "an additional stimulant" (Austen 2008a, 136) for Catherine, strengthening her determination to uncover injustice and suggesting that knowledge acquisition and recovery are efficacious medicines.

Reviving Her with Lavender Water: Recovering Sisters and Healers in *Sense and Sensibility* (1811)

Often maligned by contemporary medical texts, communities of women represented the "rich diversity and productivity of female relationships" (Williams et al. 2010, 19) and provided a means to participate in the public sphere by destabilising prescriptive gender identities (24). In *Sense and Sensibility*, Austen asserts the importance of herbal knowledge, portraying lavender as a remedy that improves nervous health and effects individual and communal recovery. The novel had its roots and was revised in the neighbourhoods of Steventon and Chawton, where women shared receipts and advice derived from lay and professional practices. Despite investigations into psychopharmacological efficacy undertaken by medical men, women continued to use efficacious plant-based remedies sourced from folk medicine. For many, the practice of domestic healthcare centred upon the household book. These multifunctional volumes, filled with culinary recipes and medical receipts, were common repositories of herbal medicine which offered alternative family histories (Leong 2013, 83; Vickery 1999, 154).

Part of a "collective endeavour" (Fara 2004, 25), Martha Lloyd's household book registers the knowledge (Leong 2013, 95) of a community and was accessible to Jane Austen during the years in which Lloyd lived with the family at Bath, Southampton and Chawton.[7] It functions as a portable collection compiled for self-sufficient, necessitous use. Receipts are entered in several hands (Black and Le Faye 2002, 34–35) and remedies contributed to combine plant-based kitchen physic with professional practice, suggesting divides and collaborations across a landscape of health care centred upon an inclusive community conversant with pharmacy (Tomalin 1998, 78; Whaley 2011, 152; Leong 2013, 95). Lloyd's household book refers to chemical substances, includes information on doses and dosing and describes how remedies might help complaints, blurring the boundaries of "scientific" (Whaley 2011, 152) and folk medicine (Lloyd, n.d., n.pag.) and lending credence to Austen's comments about Lloyd studying "medicine" at home (Austen 2011, 28). It is a manuscript significant to histories of medicine and is integral to consideration of Austen's position as a proto-feminist engaged with contemporary knowledges.

Martha Lloyd's household book includes a receipt for lavender water: "To one quart of the best rectified Spirits of wine, put three quarters of an oz; of Essence of Lavender & half a scruple of amber gris; shake it together & it is fit for use in a few days" (n.pag.). The receipt embraces hypotheses and findings on the relation between psychopharmacological efficacy and medicinal uses, such as those provided by William Lewis' *An Experimental History of the Materia Medica* (1761). As Lloyd's receipt recommends that the reader "shake" together "rectified [s]pirits of wine", "Essence of Lavender" and "amber gris", Lewis finds that lavender essential oil preserves its fragrance exactly (1784, 379) and is often produced by and/or combined with the "efficacious menstruum" (1784, xx) of rectified or "Pure" (xix) spirit of wine as the "appropriated dissolvent of resins and essential oils" (xix, xviii). Indeed, he suggests that the former, in dissolution as well as distillation (xx; 379), "extracts the virtue of lavender more completely than water" (379). Amber gris dissolves easily, allowing the essential oil to impregnate the rectified spirits and preserve the existing qualities of the remedy (45). In his discussion "Of Dissolution", John Quincy agrees that spirit of wine "easily dissolves

[7] Martha Lloyd's household book is held at Jane Austen's House, Chawton.

Resins [the primary constituent of essential oils], which *Water* can by no means do" (1719, 25).[8] The receipt's suggestion that the ingredients should be shaken together adopts contemporary ideas that a remedy's potency increased through a form of maceration (Lloyd, n.d., n.pag.; Lewis 1784, xvii; xxi). The similarities between Martha Lloyd's receipt and these publications (Rundell 1811, 329) suggest that it was shaped by the experiments and theories of "pharmaceutic chemistry" (Lewis 1784, xi). Such adaptations for medical use, with their promotion of efficacy, are evidence that women enjoyed a growing awareness of the ways in which substances, like lavender, and modifications in method and chemical process strengthened potency and improved the remedy's action and effect upon the nerves. Martha Lloyd's household book evidences women's roles in early pharmacy and the insights into efficacy, implementation and patient care shared by Austen and her circle (Whaley 2011, 156; 158; 160; Schiebinger 2004, 98).

In October 1798, Jane Austen joked to her sister, Cassandra: "I am very grand indeed; – I had the dignity of dropping out my mother's Laudanum last night" (Austen 2011, 17). Mr Lyford had "recommended" Mrs Austen "to take 12 drops of Laudanum when she went to Bed" (Austen 2011, 16). Austen updated Cassandra with detailed observations of the patient's progress: "My Mother has not been down at all today; the Laudanum made her sleep a good deal, & upon the whole I think she is better; – I shall be able to be more positive on this subject I hope tomorrow" (Austen 2011, 18). Further accounts discuss the laudanum's effects (Austen 2011, 19) and validate prognoses concerning Mrs Austen's improvement. Dispensing the prescription, Austen's observations of her mother's bodily response to the laudanum (deep and lengthy sleep that promoted recovery) evidence her awareness of the actions and effects of particular medicines. In its portrayal of the drug's potency, the letter serves as a case history that evaluates pharmaceutical efficacy (Maehle 1999, 5). Such experiences would have shaped Austen's

[8] Used by the family of Jane Austen's brother Edward, the *Knight Family Cookbook* includes a receipt originating "From Quincy's Dispensatory" (2014, 171). This citation creates a tangible connection between Quincy's catalogue, the London and Edinburgh pharmacopoeias and household books associated with Jane Austen. *The Knight Family Cookbook* [reproduced from the Knight Family Recipe Book, Hampshire: not published, 1793?], Chawton House Press, 2014 © Chawton House Library/Richard Knight.

portrayals of communities of women administering plant-based remedies that relieve nervous distress.

In *Sense and Sensibility*, Elinor Dashwood emulates Jane Austen and Martha Lloyd's convergence of folk and rational medicines. Reaching across the boundaries of perceived sexual difference and of medical practice, Elinor resembles the rational, learned woman Mary Wollstonecraft promoted in *A Vindication of the Rights of Woman* (1792). Wollstonecraft envisaged a new system of medical care that combined pre-existing nursing skills with scientific, rational medicine so that women could become equal participants in learned communities (Wollstonecraft 2008, 264). Austen's consideration of such polemic is revealed in depictions of Elinor's enfranchisement through rationality (2008b, 6) and sympathy (133), as well as effective healing and recovery (132–33; 142–43; 238). The novel's depictions of Elinor's dispensation of lavender-based preparations recall fashionable remedies and vindicate female healers by portraying as efficacious the herbal medicines chosen to alleviate nervous symptoms considered intrinsic to women's physiological and anatomical inferiority (Dyckman 2014, 232). In diagnosing Marianne's nervous distress, Elinor administers remedies approved by female healers and medical men, whose psychopharmacological knowledge of lavender's actions and effects sanctioned its continued use. Like Kitty and Catherine, Elinor uses fashionable remedies to soothe her patient, enabling Marianne's progress towards health from acute illness and functioning as Austen's means of interrogating sexual difference and its effect upon medicine.

Elinor defies theories concerning women's inherent delicacy and irrationality, serving as a mouthpiece for Austen's protest by using rationality, empirical skills and medical knowledge to recover Marianne socially and physically. While Elinor appears to respect the Palmers' apothecary, Mr Harris (2008b, 232; 234–37; Takei 2009, n.pag.), her own rationality and medical knowledge enables the evaluation of his medicines and advice. Elinor's ineffectual "proper medicines" (232) reflect the inefficacy or failure (237) of several of Mr Harris' prescriptions, suggesting that folk and rational knowledges were unstable and often equally fallible. Following her recovery, Marianne declares that Elinor is: "my nurse, my friend, my sister" (262). This recognition of Elinor's care implies that sororal affection and sympathy (Brodey 2015, 76; Knox-Shaw 2006, 148) promote healing. Elinor is an effective healer because she has shared experiences that enable her to understand Marianne and implement sympathetic care during acute nervous distress by administering efficacious lavender-based remedies. In easing Marianne's symptoms through sympathy and knowledge, Elinor embodies the psychopharmacological properties of the remedies that she administers, becoming a means of stimulating recovery and promoting the female healer's practice.

Austen's depiction of the Dashwood sisters broadcasts the importance of rational, female communities that heal physical and emotional wounds. Tony Tanner notes that Marianne is associated with sensibility, with its connections to nervous complaints (2003, 361). Yet, its effects as a cultural, political and scientific phenomenon upon her body are underpinned by Austen's refutation of sexual difference and its consequences for women (Schiebinger 1993, 38; Barker-Benfield 1992, 2; 23–28). Elinor's implementation of medicinal botany contributes to Marianne's recuperation from acute nervous symptoms, with such recovery subverting sensibility as an overpowering, often artificial, experience of feeling that disrupts women's relationships. Discovering Willoughby's presence in London, Marianne exhibits symptoms of acute nervous distress: "Marianne, now looking dreadfully white, and unable to stand, sunk into her chair, and Elinor, expecting every moment to see her faint, tried to screen her from the observation of others, while reviving her with lavendar [sic] water" (132). Elinor observes Marianne's sudden blanching, interpreting it as a symptom of the heightened sensibility or emotional responses indicative of weakened nerves or "[f]ibres" (Cheyne 1733, 102) associated with nervous disease (Porter 1987, 180–81; Gorman 1993, 139; 142–43; Buchan 1798, 443). Fainting fits were thought to be physiological responses caused by external stimuli, like fear (Buchan 2014, 554; Darwin 2009, 2:I.ii.i.iv 67; Brown 1795, 1:13). Consequently, Marianne's grief acts as a stimulus that undermines her nervous health and provokes her near-faint, indicating the exhaustion of her nervous system and vital functions.

Marianne's paleness and faintness convey asthenic overstimulation that culminates in nervous distress (Kaplan and Kaplan 1990, 117–30). Gavin Budge explains that asthenia "was understood as a condition of addiction to stimulation" (2013, 56), where repeated use of stimuli could render individuals debilitated. Kimiyo Ogawa has argued that Marianne's attachment to Willoughby becomes an addiction to sex and sensibility that is remedied by Elinor's influence (2012, n.pag.). Intoxicating Marianne through enthusiastic recognition of the importance of their attachment, Willoughby becomes a stimulus whose effect on Marianne's passions

(Beddoes 1795, 1:clviii) causes debility that becomes evident following their estrangement. Elinor implements herbal knowledge, improving Marianne's asthenic condition through nervous exertion (Darwin 2009, 1:XII.vii.i 96) and countering Willoughby's debilitating effect by using lavender, sororal love and comfort (Budge 2013, 56–57; Darwin 2009, 1:XII.vii.iii 96; Brown 1795, 1:13).

In his description of the "Cephalicks", or "Cordials", that revived the nervous system from the "Languor" (Quincy 1719, 77) responsible for faintness, Quincy detailed the ways in which "Nervous Simples" (76) like lavender (82) improved physiological processes and effected recovery (77–78). His consideration of "volatile" substances added to smelling-bottles traced their journey between the nerves and organs (78), assessing efficacy through hypothesis dependent upon observation of medicinal use and patient need. Marianne's visible paleness and debility shape Elinor's response to her acute nervous distress (Beddoes 1795, 1:clv–clviii). By "reviving" Marianne "with lavendar [sic] water" (Austen 2008b, 132), Elinor administers an "emergency remedy" (Friedman 2016, 56) that soothes her sister and supports her improvement. Taken externally through inhalation, its "reviving smells" (Friedman 2016, 55) effected an immediate recovery, while its smelling-bottle symbolised the user's experience of feeling and disease (Friedman 2016, 54; Beatty 2012, 66). Yet, use of lavender water advocates privacy, "screen[ing]" (Austen 2008b, 132) Marianne and concealing the sisters' upset as it enables Elinor's informed observation of its soothing, revivifying effects. Thus, Elinor uses a remedy sanctioned for acute nervous distress but divests it of fashionable connotations, actively administering the lavender water to preserve Marianne's dignity, agency and nervous health. Although she remains in a state of "agitation", with focus immoveable from Willoughby, Elinor's lavender water regains Marianne's articulacy and mobility (132).

Austen's depiction of Elinor's conscientious, efficacious care asserts the psychopharmacological viability of fashionable remedies and vindicates women's herbal knowledge. Dispensing a plant-based substance popular in folk and rational medicines, Elinor follows advice concerning the patient's recovery from nervous distress and exhibits an awareness of lavender's hypothetical actions upon the body and visible effects by alleviation of her sister's distress. Regarded as a "strengthening medicine" (Beatty 2012, 118–22), Cullen stated that lavender "contains a large portion of essential oil of a very grateful odour. It is therefore, whether externally applied or given internally, a powerful stimulant to the nervous system" (1789, 2:147). Austen's portrayals of Elinor's treatment of Marianne's distress are influenced by pharmacopoeias and advice manuals (Buchan 1798, 443), the lavender water's inhalation acting by stimulating Marianne's brain and nerves and subduing her noticeable distress. Evinced through its efficacy, Elinor's knowledge of lavender's actions and effects upon the nervous system supports a reading of Austen as a learned woman conversant with scientific hypotheses surrounding the nerves and the ways in which corresponding herbal treatments might improve nervous health. Elinor's knowledge of potent plant-based remedies used to alleviate nervous distress recovers Marianne, revealing an image of a female healer enfranchised by successful healing.

Following the receipt of Willoughby's last letter, Marianne grows increasingly distressed. Elinor dispenses lavender drops to ease the symptoms:

> Elinor advised her to lie down again, and for a moment she did so; but no attitude could give her ease; and in restless pain of mind and body she moved from one posture to another, till growing more and more hysterical, her sister could with difficulty keep her on the bed at all, and for some time was fearful of being constrained to call for assistance. Some lavender drops, however, which she was at length persuaded to take, were of use; and from that time till Mrs. Jennings returned, she continued on the bed quiet and motionless. (Austen 2008b, 142–43)

Austen's allusion to the drops is underpinned by rational advice that considered them an efficacious remedy. Elinor's lavender, or "palsy", drops would have been comprised of "compound spirits of lavender" (Buchan 2014, 501) and were ingested to relieve symptoms attending nervous disease. With their ability to regulate the nerves, lavender drops were considered an effective remedy for hysteria (Lewis 1784, 380; Quincy 1719, 82). Buchan declared: "NERVOUS disorders often proceed from affections of the mind, as grief, disappointments, anxiety, intense study, &c" (2014, 509). Such stimuli could cause hysteria, its asthenic symptoms exhibited through "mobility and changeableness of mind, disposition to sleep [and a] convulsive state" similar to palsy (Brown 1795, 2:262). Prompted by Marianne's bereft expressions and erratic movements, Elinor administers the drops because she concludes that her sister's acute distress is symptomatic of chronic nervous disease, due to its "hysterical" quality (Austen 2008b, 143). Following Brunonian advice on "mild hysteria", she implements "a moderate and naturally stimulant management" (Brown 1795, 2:230). Marianne's response to grief is a harbinger of her later illness, but her recovery from acute nervous distress foreshadows Elinor's success as a healer. Cullen outlines the physiological effects of lavender on the nerves, explaining that it gently excites the "brain to a fuller impulse of the nervous power into the nerves of the animal functions" (1789, 2:148). When inhaled or ingested, lavender acts upon the nerves, strengthening them and improving the patient's response to external stimuli. Consequently, Elinor administers preparations according to their action upon the nerves and composing, soothing effects upon distress. Marianne's continuing "quiet and motionless" (Austen 2008b, 143) suggests that she has fallen asleep or unconscious, evidencing that the lavender drops are efficacious. Austen's descriptions of Elinor observing her sister's distress, as well as the lavender's action and effects, and of Marianne ingesting and responding to the drops, reflects contemporary, professional methods of assessing the psychopharmacological efficacy of items included in pharmacopoeias, such as "observing the effects" on animals following injection of drug substances (Cullen 1789, 1:153–54). The results of experiments involving ingestion were regarded as unreliable by certain medical men, but Elinor's ability to evaluate the effects of all medicines dispensed to her sister asserts the female healer's ability to administer and investigate remedies associated with lay and professional knowledge and to manage health care through folk and rational medicines.

Observation and experimentation characterise Austen's writings. All interrogate scientific discourses and invite the reader to observe the effects of the heroine's medical knowledge and practice, investigating the efficacy and value of medicinal botany and the aetiology of women's marginalisation. In Austen's novella and Steventon novels, lavender is carried and used as a means of protest by young women who attempt to heal and to reclaim control of domestic health care through remedies which overcome acute nervous distress. Austen criticises the ways in which theories of sexual difference affected women's experience of nervous symptoms and of the plant-based remedies used in their treatment. The depiction of lavender water's delayed efficacy in *Northanger Abbey*, for example, allows Austen to fictionalise the turbulence within proto-pharmacological scrutiny of folk remedies and its implications for the female healer's practice. In "Kitty", such restrictions upon women are challenged through the heroine's wielding of botanical knowledges and insistence upon the continued relevance of female friendship, as well as folk remedies, to women's enfranchisement. These attempts at recovery are crystallised in *Sense and Sensibility*, where Elinor Dashwood utilises folk and rational knowledges and practices, with their nursing and empirical skills, to assert lavender's pharmaceutical efficacy and to recover her sister's social and nervous health. Consequently, Austen's allusions to lavender become a means of restoring herbal medicine to proto-feminist communities through the healer's psychopharmacological knowledge and efficacious medical care of female patients.

References

Aikin, John. 1784. Advertisement. In *An Experimental History of the Materia Medica, Or of the Natural and Artificial Substances Made Use of in Medicine*. By William Lewis, ed. John Aikin. London: J. Johnson.

Anonymous. 2014. *The Knight Family Cookbook*. Alton: Chawton House Press.

Austen, Jane. 1792. *Volume the Third*. MS. Add MS 65381. British Library, London.

———. 1951. *Volume the Third*, ed. R. W. Chapman. Oxford: Clarendon Press.

———. 1963. *The Works of Jane Austen. Volume VI: Minor Works*, ed. R. W. Chapman, 6 vols. London: Oxford University Press.

———. 2008a. *Northanger Abbey, Lady Susan, the Watsons and Sanditon*, ed. James Kinsley and John Davie. Oxford: Oxford University Press.

———. 2008b. *Sense and Sensibility*, ed. James Kinsley. Oxford: Oxford University Press.

———. 2009. *Catharine and Other Writings*, ed. Margaret Anne Doody and Douglas Murray. Oxford: Oxford University Press.

———. 2011. *Jane Austen's Letters*, ed. Deirdre Le Faye. Oxford: Oxford University Press.

———. 2013. *Jane Austen's Manuscript Works*, ed. Linda Bree, Peter Sabor, and Janet Todd. London: Broadview Press.

———. 2017. *Teenage Writings*, ed. Kathryn Sutherland and Freya Johnston. Oxford: Oxford University Press.

Barker-Benfield, G.J. 1992. *The Culture of Sensibility: Sex and Society in Eighteenth-Century Britain*. Chicago: University of Chicago Press.

Beatty, Heather R. 2012. *Nervous Disease in Late Eighteenth-Century Britain: The Reality of a Fashionable Disorder*. Abingdon: Routledge.

Beddoes, Thomas. 1795. Preface. In *The Elements of Medicine of John Brown M.D.*, ed. and trans. Thomas Beddoes. London: J. Johnson.

Black, Maggie, and Deirdre Le Faye. 2002. *The Jane Austen Cookbook*. London: British Museum; Toronto: McLelland and Stuart.

Blackwell, Elizabeth. 1737–39. *A Curious Herbal*, 2 vols. London.
Boerhaave, Herman. 1755. *Herman Boerhaave's Materia Medica, or the Druggist's Guide, and the Physician and Apothecary's Table-Book*. London.
Boorman, William Harry. 1966. Jane Austen & 'Mr. Lyford'. TS. 120M94W/C80. Hampshire Record Office, Winchester.
Brodey, Inger Sigrun. 2015. Making Sense of Sensibility. *Persuasions* 37: 62–80.
Brown, John. 1795. *The Elements of Medicine of John Brown M.D.*, ed. and trans. Thomas Beddoes, 2 vols. London: J. Johnson.
Buchan, William. 1798. *Domestic Medicine: Or, a Treatise on the Prevention and Cure of Diseases by Regimen and Simple Medicines*. London.
———. 2014 [1769]. *Domestic Medicine: Or, the Family Physician*. Cambridge: Cambridge University Press.
Budge, Gavin. 2013. *Romanticism, Medicine and the Natural Supernatural: Transcendent Vision and Bodily Spectres, 1789–1852*. Basingstoke: Palgrave Macmillan.
Cheyne, George. 1733. *The English Malady*. London.
Chute, Eliza. 1797. *Diary*. MS. 23M93/70/1/5. Austen-Leigh Archive, Hampshire Record Office, Winchester.
Cohen, Michèle. 2006. 'A Little Learning'? The Curriculum and the Construction of Gender Difference in the Long Eighteenth Century. *British Journal for Eighteenth-Century Studies* 29: 321–35.
Cullen, William. 1789. *A Treatise of the Materia Medica*. 2 vols. Edinburgh.
Darwin, Erasmus. 2009 [1794-96]. *Zoonomia; or the Laws of Organic Life*, 2 vols. Cambridge: Cambridge University Press.
Dolan, Elizabeth A. 2008. *Seeing Suffering in Women's Literature of the Romantic Period*. Aldershot: Ashgate.
Dyckman, Ivy. 2014. You've Got the Vapors?: Historical, Medical and Literary Perspectives. In *Women, Gender and Disease in Eighteenth-Century England and France*, ed. Kathleen Hardesty Doig and Felicia Berger Sturzer, 231–52. Newcastle upon Tyne: Cambridge Scholars.
Fara, Patricia. 2004. *Pandora's Breeches: Women, Science and Power in the Enlightenment*. London: Random House.
Faubert, Michelle. 2005. A Gendered Affliction: Women, Writing, Madness. In *Cultural Constructions of Madness in Eighteenth-Century Writing: Representing the Insane*, ed. Allan Ingram, 136–69. Basingstoke: Palgrave Macmillan.
Francia, Susan, and Anne Stobart. 2014. The Fragmentation of Herbal History: The Way Forward. In *Critical Approaches to the History of Western Herbal Medicine: From Classical Antiquity to the Early Modern Period*, ed. Susan Francia and Anne Stobart, 1–20. London and New York: Bloomsbury.
Friedman, Emily C. 2016. *Reading Smell in Eighteenth-Century Fiction*. Lanham, MD: Bucknell University Press.

Gorman, Anita G. 1993. *The Body in Illness and Health: Themes and Images in Jane Austen*. New York: Peter Lang.

Hill Curth, Louise. 2006. Medical Advertising in the Popular Press: Almanacs and the Growth of Proprietary Medicines. In *From Physick to Pharmacology: Five Hundred Years of British Drug Retailing*, ed. Louise Hill Curth, 29–47. Aldershot: Ashgate.

Kaplan, Laurie, and Richard S. Kaplan. 1990. What's Wrong with Marianne? Medicine and Disease in Jane Austen's England. *Persuasions* 12: 117–30.

King, Steven. 2006. Accessing Drugs in the Eighteenth-Century Regions. In *From Physick to Pharmacology: Five Hundred Years of British Drug Retailing*, ed. Louise Hill Curth, 49–78. Aldershot: Ashgate.

Knox-Shaw, Peter. 2006. *Jane Austen and the Enlightenment*. Cambridge: Cambridge University Press.

Lastinger, Valérie. 2014. The Laboratory, the Boudoir and the Kitchen: Medicine, Home and Domesticity. In *Women, Gender and Disease in Eighteenth-Century England and France*, ed. Kathleen Hardesty Doig and Felicia Berger Sturzer, 119–48. Newcastle upon Tyne: Cambridge Scholars.

Leong, Elaine. 2013. Collecting Knowledge for the Family: Recipes, Gender and Practical Knowledge in the Early Modern English Household. *Centaurus* 55 (2): 81–103.

Lewis, William. 1784. *An Experimental History of the Materia Medica, Or of the Natural and Artificial Substances Made Use of in Medicine*, ed. John Aikin. London: J. Johnson and R. Baldwin.

Lloyd, Martha. n.d. *Household Book.* MS. CHWJA:JAHB34. Jane Austen's House, Chawton.

Loudon, Irvine. 1986. *Medical Care and the General Practitioner, 1750–1850.* Oxford: Oxford University Press.

Lynch, Deidre Shauna. 2010. 'Young Ladies Are Delicate Plants': Jane Austen and Greenhouse Romanticism. *ELH* 77: 689–729.

Maehle, Andreas-Holger. 1999. *Drugs on Trial: Experimental Pharmacology and Therapeutic Innovation in the Eighteenth Century*. Amsterdam and Atlanta, GA: Rodopi.

Ogawa, Kimiyo. 2012. Marianne's Addiction: Amorous Pleasures in *Sense and Sensibility*. *Persuasions On-Line*. 32 (2). http://www.jasna.org/persuasions/on-line/vol32no2/ogawa.html. Accessed 1 February 2020.

Page, Judith W., and Elise L. Smith. 2011. *Women, Literature, and the Domestic Landscape: England's Disciples of Flora, 1780–1830*. Cambridge: Cambridge University Press.

Porter, Roy. 1987. *Mind-Forg'd Manacles*. London: Athlone Press.

———. 1992. The Patient in England, c.1660–c.1800. In *Medicine in Society: Historical Essays*, ed. Andrew Wear, 91–118. Cambridge: Cambridge University Press.

Quincy, John. 1719. *Pharmacopœia Officialis & Extemporanea: Or, A Compleat English Dispensatory, In Four Parts.*

Raffald, Elizabeth. 1788. *The Experienced English Housekeeper, for the Use and Ease of Ladies, Housekeepers, Cooks, &c.* London.

Royal College of Physicians of Edinburgh. 1744. *Pharmacopoeia Collegii Regii Medicorum Edinburgensis.*

Royal College of Physicians of London. 1746. *Pharmacopoeia Collegii Regalis Medicorum Londinensis.*

Rundell, Maria. 1811. *The New Family Receipt Book, Containing Eight Hundred Truly Valuable Receipts in Various Branches of Domestic Economy.* London: John Murray.

Sabor, Peter. 2018. *Reading with Austen.* Burney Centre, McGill University; Chawton House. www.readingwithausten.com. Accessed 1 February 2020.

Schiebinger, Londa. 1993. *Nature's Body: Gender in the Making of Modern Science.* Brunswick: Rutgers University Press.

———. 2004. *Plants and Empire: Colonial Bioprospecting in the Atlantic World.* Cambridge and London: Harvard University Press.

———. 1991. *The Mind Has No Sex?: Women in the Origins of Modern Science.* Cambridge and London: Harvard University Press.

Shteir, Ann B. 1996. *Cultivating Women, Cultivating Science: Flora's Daughters and Botany in England, 1760–1860.* Baltimore: Johns Hopkins University Press.

Smith, Ginnie. 1985. Prescribing the Rules of Health: Self-Help and Advice in the Late Eighteenth Century. In *Patients and Practitioners: Lay Perceptions of Medicine in Pre-Industrial Society*, ed. Roy Porter, 249–82. Cambridge: Cambridge University Press.

Stobart, Anne. 2016. *Household Medicine in Seventeenth-Century England.* London: Bloomsbury.

Stone, Lawrence. 1990. *The Family, Sex and Marriage in England 1500–1800.* London: Penguin.

Sutherland, Kathryn. 2010. *Jane Austen's Fiction Manuscripts: A Digital Edition.* AHRC. http://www.janeausten.ac.uk/facsimile/blvolthird/70.html. Accessed 1 February 2020.

———. 2015. From Kitty to Catharine: James Edward Austen's Hand in *Volume the Third. Review of English Studies* 66 (273): 124–43.

Sutherland, Kathryn, and Freya Johnston. 2017a. Introduction. In *Teenage Writings.* By Jane Austen, ed. Kathryn Sutherland and Freya Johnston. Oxford: Oxford University Press.

———. 2017b. Textual Notes. In *Teenage Writings.* By Jane Austen, ed. Kathryn Sutherland and Freya Johnston. Oxford: Oxford University Press.

Takei, Akiko. 2009. 'Mr. Cole Is Very Bilious': The Art of Lay Medicine in Jane Austen's Characters. *Persuasions On-Line* 30 (1). http://jasna.org/persuasions/on-line/vol30no1/takei.html?. Accessed 1 February 2020.

Tanner, Tony. 2003. Appendix: Original Penguin Classics Introduction. In *Sense and Sensibility*. By Jane Austen, ed. Ros Ballaster. London: Penguin.

Tomalin, Claire. 1998. *Jane Austen: A Life*. London: Penguin.

Tuite, Clara. 2002. *Romantic Austen: Sexual Politics and the Literary Canon*. Cambridge: Cambridge University Press.

Tullett, William. 2019. *Smell in Eighteenth-Century England: A Social Sense*. Oxford: Oxford University Press.

Vickery, Amanda. 1999. *The Gentleman's Daughter: Women's Lives in Georgian England*. New Haven and London: Yale University Press.

Whaley, Leigh. 2011. *Women and the Practice of Medical Care in Early Modern Europe, 1400–1800*. Basingstoke: Palgrave Macmillan.

Williams, Carolyn D., Angela Escott, and Louise Duckling. 2010. Introduction. In *Woman to Woman: Female Negotiations During the Long Eighteenth Century*, ed. Carolyn D. Williams, Angela Escott, and Louise Duckling, 19–45. Newark: University of Delaware Press.

Wiltshire, John. 2005. Medicine, Illness and Disease. In *Jane Austen in Context*, ed. Janet Todd, 306–16. Cambridge: Cambridge University Press.

Wollstonecraft, Mary. 2008. *A Vindication of the Rights of Woman* and *A Vindication of the Rights of Men*, ed. Janet Todd. Oxford: Oxford University Press.

———. 2009. *Mary* and *The Wrongs of Woman*, ed. Gary Kelly. Oxford: Oxford University Press.

Worth Estes, J. 1997. The Therapeutic Crisis of the Eighteenth Century. In *The Inside Story of Medicines: A Symposium*, ed. Gregory J. Higby and Elaine C. Stroud. Madison: American Institute of the History of Pharmacy.

Zieger, Susan. 2008. *Inventing the Addict: Drugs, Race, and Sexuality in Nineteenth-Century British and American Literature*. Amherst: University of Massachusetts Press.

CHAPTER 12

"When Poor Mama Long Restless Lies, / She Drinks the Poppy's Juice": Opium and Gender in British Romantic Literature

Joseph Crawford

Opium played a double role in British society during the Romantic period. It could be bought over the counter throughout Britain, with no restrictions placed upon its sale until the Pharmacy Act of 1868 (Berridge 1999, 21). And yet, when the traveller and physician Richard Madden visited Turkey in 1824, one of the first things he did was to visit a coffeehouse frequented by Turkish "*Theriakis*, or opium eaters" (Madden 1833, 18), eager to explore such an exotic and unfamiliar scene of "voluptuousness" and to report back on his experiences for the benefit of his fascinated correspondents back home. For Madden, the *Theriakis* of Constantinople were "frightful" and decidedly foreign figures, dreaming of "harems" and "celestial *houris*", and enslaved to a drug so destructive that they seldom "[pass] thirty years of age" (Madden 1833, 18–21). Yet the drug they took, far from being a mysterious product of the distant East, was one that could be purchased in every village in Britain, and

J. Crawford (✉)
University of Exeter, Exeter, UK
e-mail: J.Crawford@exeter.ac.uk

N. Roxburgh and J. S. Henke (eds.), *Psychopharmacology in British Literature and Culture, 1780–1900*, Palgrave Studies in Literature, Science and Medicine,
https://doi.org/10.1007/978-3-030-53598-8_12

which the British consumed in such quantities that, by 1838, the nation was importing over 130,000 pounds of opium for domestic consumption every year (Porter and Porter 1988, 219–20; Foxcroft 2007, 95). How could the same drug be simultaneously so foreign and so commonplace, its use both deviant and utterly mundane?

For Madden and his contemporaries, what distinguished the *Theriakis* from the majority of British opium users was the fact that they took opium for pleasure rather than for what were then regarded as legitimate medicinal reasons. The idea that anyone taking regular heavy doses of opium should be classed as suffering from a pathology called "opium addiction", regardless of their reasons for taking it, would have seemed strange to most British doctors of the period: and just as the early British temperance activists campaigned against over-indulgence in gin while having no objection to the regular consumption of beer, so most doctors saw nothing wrong with the "moderate employment" of opiates, even when taken on a regular basis for years on end (Berridge 1999, 86; Shiman 1988, 9–15). With the concept of addiction still in its infancy, British opium users were judged not on the basis of whether they were dependent upon the drug, but in terms of whether they took it as a stimulant or a relaxant, and whether their reasons for doing so were moral or immoral, selfish or selfless (Zieger 2008, 4–5).

Although opium is now classed as a depressant, in Madden's day there was great disagreement about whether the drug should be considered primarily as a stimulant or a sedative (Risse 1988, 50). As Dormandy notes, by the later eighteenth century it had become clear that opium could function either as a "stimulant" or as a "relaxant": "Depending on dosage and, even more, on striking differences between individual responses, it was in fact – or could be – both or neither", as "opium in any dose could both soothe and conjure up strange and exhilarating new ideas" (Dormandy 2012, 60; 82). These different notions of drug effect, however, did not have equal cultural status. The use of opium as a sedative and painkiller was routine and unremarkable, but the recreational use of the drug to stimulate the mind and body was the subject of widespread disapproval and was associated with the deviant, degenerate figure of the "oriental" opium-eater (Berridge 1999, 28–33; 53–54; Foxcroft 2007, 61–63). Thus George Young's influential *Treatise on Opium* (1753) described opium as a "soporific" drug which was "chiefly useful, as it abates a *stimulus*", but he contrasted his own medical use of opium with that of the Turks, who "take opium freely", remarking that "the Turks doubt no more of its power to exhilarate and give courage, than we do of its soporific quality" (Young 1753, 14; 61; 104).

Thomas Trotter's *Essay on Drunkenness* (1804, 1813) classed both opium and alcohol as "narcotics" causing "stupour and sleep", though he noted that there was great disagreement over whether opium was a sedative or a stimulant that merely had a secondary sedative effect due to the exhaustion which inevitably succeeded the excitement it provoked (Trotter 1813, 40). Trotter himself, like Young before him, argued the fact that "Turks and East Indians" used opiates as "stimulants and cordials" demonstrated that their effects were not purely sedative, but he made clear in both his *Essay* and his later *View of the Nervous Temperament* (1807) that he strongly disapproved of opium being used in this way. In his view, the only legitimate medical use of opium was "as an anodyne", as a last-ditch treatment for "pain and restlessness" when other remedies proved ineffectual (Trotter 1807, 136–41; 312; 317–19; 322; 1813, 42; 44; 46). In a similar vein, John Reid's *Essays on Hypochondriacal and Other Nervous Affections* (1816) acknowledged that opium could be used as either a stimulant or a sedative but argued that only the latter usage was medically legitimate. Like Trotter, Reid was willing to sanction the use of opium as an emergency painkiller, but he warned that only the horrors of withdrawal awaited those who used opium as an intoxicant (Reid 1816, 87–91; 100–06).

Not all contemporary doctors disapproved of the stimulating powers of opium. The heterodox medical theorist John Brown regarded opium, like alcohol, as a stimulant, and on this basis he prescribed it freely to patients whom he regarded as suffering from "asthenia", or lack of vital energy. Notoriously, Brown used these same treatments to stimulate his own vital energies, ultimately dying of an opium overdose in 1788 (Budge 2012, 55–56; Lawrence 1988, 5; Youngquist 1999, 156; Grinnell 2006, 237). Brown's followers, who included Thomas Beddoes and Erasmus Darwin, continued to prescribe opium as a stimulant, although bitter experience ultimately led some of them to modify their practice after discovering that its supposedly life-stimulating powers were useless when treating infectious diseases, and that large doses risked killing patients rather than curing them (Darwin 1794, 71; 364; Vickers 1997, 65–66; 72; Kondratas 1988, 80–81). Even Brunonians, however, warned against the recreational use of opium as a stimulant by otherwise healthy individuals, as according to their paradigm the inappropriate use of opiates risked placing such people in a "sthenic" state just as dangerous as the "asthenia" which opium was more properly used to cure. As Erasmus Darwin wrote, "the art of preserving long health and life [...] must consist in using no greater stimulus [...] than is sufficient to preserve us in vigour". Using opium when it was not medically necessary thus risked wearing out the body, condemning the user to an early death (Darwin 1794, 468–69).

The two most famous "deviant" users of opium in the Romantic period were Samuel Taylor Coleridge and Thomas De Quincey, both of whom were repeatedly accused, not least by one another, of having taken opium "as a *stimulant*", or "purely for the sake of creating an artificial state of pleasurable excitement" (Foxcroft 2007, 25–26; Leask 1992, 183–85; Berridge 1999, 51–52).[1] Even though both Coleridge and De Quincey primarily used the drug for its sedative and analgesic properties, their writings helped to create a stereotype which, as Leask notes, was firmly established by the end of the Romantic period: that of the opium abuser as a morally irresponsible visionary, who seeks inspiration in drug-induced dreams of fantastical Asiatic landscapes. Leask writes:

> Treatments of drug experience in the period drew on a consistent range of stereotypes: the hedonistic – or in newspeak 'recreational' – use of drugs revealed a quasi-mystical site of truth in the psyche, in many cases a vessel for spilt religion or repressed sexuality and creativity. This was balanced by an anxiety about the moral consequences of instant pharmaceutical gratification [...] and paranoiac visions of alien, exotic – usually oriental – cultures from the beginning associated with western narcosis. (Leask 1992, 171)

What Leask describes here was the contemporary stereotype of the hedonistic opium user as a countercultural figure, his mind and body infected by foreign drugs. I use the male pronoun deliberately, for such individuals were usually imagined as male, just as Madden saw only men among the *Theriakis* of Constantinople. It was male writers such as Branwell Brontë and William Blair who would later reflect, ruefully, on how De Quincey had inspired them to begin using opium (Berridge 1999, 57; Blair 1842, 49). When John Wilson alluded to the idea that De Quincey's *Confessions of an English Opium-Eater* had caused a rash of accidental

[1] In this they formed a striking contrast to the medically trained John Keats, who consistently emphasised the drug's sedative powers in his poetry (Keats 1970, 70; 511; 525; 652; 660–61). See also essay in this volume by Octavia Cox.

suicides, he wrote in his "Noctes Ambrosianae", "What if fifty foolish fellows have been buried in consequence of that delightful little Tractate of Education?", clearly assuming that it would be men rather than women who were likely to end up unintentionally overdosing on opium in the hope of experiencing the kind of visions that De Quincey had described (Wilson 1843, 1 226).

A number of scholars, including Abrams (1934), Hayter (1968), Leask (1992), Berridge (1999), Foxcroft (2007), and Dormandy (2012), have written on the importance of opium to the writings of Coleridge and De Quincey, and of the later male writers who imitated them. It has seldom been noted, however, that another literature of opium use existed in Romantic-era Britain, which is to be found in the less well-known works of their female contemporaries. This women's opium literature depicted the drug in a rather different light: not as a seductive and destructive stimulant which enticed the (male) artist away from his domestic responsibilities into exotic "oriental" vistas, but as a form of sedative self-medication which helped its (female) users to cope with the demands imposed by the very world of mundane domesticity which male opium-eaters used the drug to forsake. In this chapter, I shall argue that this distinction between male and female writing about opium in the period corresponds to the double role played by the drug in contemporary British society as a whole. By aligning their use of opium with the "moderate" and medicinal consumption of opiates recommended by contemporary doctors rather than with the deviant figure of the selfish and hedonistic opium-eater, these female writers were able to imply that their drug use was compatible with, and indeed contributed to, their ability to selflessly and effectively discharge their domestic duties (Berridge 1999, 86).

It is tempting to draw connections between Coleridge's opium habit and the phantasmagorical quality of his poetic imagination. Coleridge himself hinted at such possibilities in his preface to "Kubla Khan", in which he claimed that this "psychological curiosity" of a poem was inspired by a drug-induced dream:

> In the summer of the year 1797, the Author, then in ill health, had retired to a lonely farm-house between Porlock and Linton, on the Exmoor confines of Somerset and Devonshire. In consequence of a slight indisposition, an anodyne had been prescribed, from the effects of which he fell asleep [...][.] The Author continued for about three hours in a profound

> sleep, at least of the external senses, during which time he has the most vivid confidence, that he could not have composed less than from two to three hundred lines; if that indeed can be called composition in which all the images rose up before him as *things*, with a parallel production of the correspondent expressions, without any sensation or consciousness of effort. On awaking he appeared to himself to have a distinct recollection of the whole, and taking his pen, ink, and paper, instantly and eagerly wrote down the lines that are here preserved. At this moment he was unfortunately called out by a person on business from Porlock, and detained by him above an hour. (Coleridge 1969, 295–96)

Coleridge's "anodyne" was, of course, opium, and "Kubla Khan" has often been interpreted as a poetic evocation of an opium-vision.[2] In a similar vein, De Quincey's 1821 *Confessions* described some of the extraordinary dreams which he experienced under the influence of opium:

> I was stared at, hooted at, grinned at, chattered at, by monkeys, by paroquets, by cockatoos. I ran into pagodas: and was fixed, for centuries, at the summit, or in secret rooms: I was the idol; I was the priest; I was worshipped; I was sacrificed. I fled from the wrath of Brama through all the forests of Asia: Vishnu hated me: Seeva laid wait for me. [...] The cursed crocodile became to me the object of more horror than almost all the rest. I was compelled to live with him, and (as was always the case almost in my dreams) for centuries. [...] And so often did this hideous reptile haunt my dreams that many times the very same dream was broken up in the very same way: I heard gentle voices speaking to me (I hear everything when I am sleeping); and instantly I awoke: it was broad noon; and my children were standing, hand in hand, at my bedside; come to show me their coloured shoes, or new frocks, or to let me see them dressed for going out. (De Quincey 1986, 109–10)

While the emotional tenor of De Quincey's nightmare is very different to that of Coleridge's exultant vision, both have several features in common. In each case, a lone male writer describes their visionary experience under the influence of opium, in which they imagine "oriental" vistas—Coleridge's Xanadu, De Quincey's "forests of Asia"—filled with mysterious and supernatural scenes, far removed from their everyday

[2] The 2001 film *Pandemonium* went further, depicting both "Kubla Khan" and "The Rime of the Ancient Mariner" as being more-or-less direct transcriptions of Coleridge's opium dreams.

English lives. They are completely swept up in their visions, within which time seems to dilate: Coleridge composes three hundred lines in three hours, while De Quincey experiences subjective eternities in a single night. Finally, they are dragged out of their visions—De Quincey gratefully and Coleridge reluctantly—by the arrival of figures who stand for the everyday business of normal life: De Quincey by his children and Coleridge by a "person on business from Porlock".

Coleridge and De Quincey singled out these dream-visions for description specifically because of their unusual intensity, and most of the occasions on which they took opium were attended by much less spectacular drug effects. However, they became key parts of the mythology of opium abuse that swiftly developed around both writers—a mythology which aligned hedonistic opium usage with the foreign, the "oriental" and the wondrous, all of which stand in stark contrast to the domestic realities which await these writers when their dream-visions finally end. A similar sense of conflict between hedonistic male opium reverie and the domestic world can be seen in Maria Brooks' 1820 poem "TO ONE, who had taken laudanum to enliven himself", in which a male opium addict seeks "Mad pleasure's throb" by using opium as a recreational stimulant, alienating himself from the poem's female speaker in the process. Given the gendered dynamics at work in this scene, it is telling that she attempts to persuade him to forsake the drug by encouraging him to think of his mother (Brooks 1820, 29–31).

The contemporary perception of Coleridge and De Quincey as paradigmatic examples of "deviant" opium users is clear from an article in the 1822 *Quarterly Journal of Foreign and British Medicine and Surgery*, in which De Quincey's *Confessions* was reviewed side-by-side with *On the Nature of Opium*, a medical treatise by the Viennese doctor Christopher Christen. After spending several pages seriously discussing Christen's views on the medical uses of opium in treating various illnesses, as well as its advantages and disadvantages as a surgical anaesthetic and the potential benefits of cultivating opium poppies in Britain on a commercial scale, the reviewer turns to De Quincey's *Confessions* with contempt: "The story itself, though it possesses some interest at the commencement, is both ill-told and aimless; and we believe, that few have had patience to read beyond the introduction, which, by the way, is full of the gross and disgusting immorality of the German school, that produced Werter [sic], and the Stranger" (Anonymous 1822, 362). Having castigated De Quincey for his immorality, the reviewer then quotes the same passage from the *Confessions* cited above as an example of his "absolute raving", before going on to deliver his ultimate put-down: the *Confessions* are such "nonsense" that they "bring strongly to our recollection certain dreams and ravings of Mr Coleridge", who is here described as "a stray poet, stumbling among the mountains of Cumberland, bewrapt in an opium reverie" (Anonymous 1822, 362–63). The medical use of opium by respectable doctors such as Christen, who use it as a sedative to combat "excessive sensibility or excitement", is thus contrasted with the recreational opium abuse of Coleridge and De Quincey, whose hyperactive "raving" and "stumbling" imply that they have been using opium as a stimulant rather than as a relaxant, and who are described in suggestively orientalising terms as "*running a muck*, at morality, or medicine" (Anonymous 1822, 355; 363). As Zieger notes, the custom of running amok was believed, at the time, to be caused by the fact that the people of Malaysia combined Asiatic "savagery" with opium abuse (Zieger 2008, 41). By applying this phrase to Coleridge and De Quincey, the reviewer thus suggests that both men have been so corrupted by their "oriental" habit of hedonistic opium-eating that they are now in danger of running amok like crazed Malays (Anonymous 1822, 363).

Coleridge's opium addiction has become famous, but it is much less well known that Coleridge's daughter, Sara, also became an opium addict. Like many British women of the period, she took opium habitually as a painkiller and sedative, and gradually became dependent upon it (Foxcroft 2007, 44). She described her opium usage in her poem, "Poppies":

> The poppies blooming all around
> My Herbert loves to see;
> Some pearly white, some dark as night,
> Some red as cramasie:
>
> He loves their colours fresh and fine,
> As fair as fair may be;
> But little does my darling know
> How good they are to me.
>
> He views their clust'ring petals gay,
> And shakes their nut-brown seeds;
> But they to him are nothing more

Than other brilliant weeds.

O! how shouldst thou, with beaming brow,
With eye and cheek so bright,
Know aught of that gay blossom's power,
Or sorrows of the night?

When poor Mama long restless lies,
She drinks the poppy's juice;
That liquor soon can close her eyes,
And slumber soon produce:

O then my sweet, my happy boy
Will thank the Poppy-flower,
Which brings the sleep to dear Mama,
At midnight's darksome hour. (Feldman 1997, 199)

In this poem, the use of opium is not contrasted with domestic life, but embedded within it. Because the speaker is Herbert's "dear Mama", she *has* to be able to sleep, in order to carry out her domestic duties as a mother. When she "long restless lies", tormented by "sorrows of the night", the use of opium becomes a necessity, enabling her to attain the "slumber" that, as the mother of a young child, she so desperately requires. But rather than presenting the "happy boy" and the "liquor" of the opium poppy as opposing forces, this poem suggests a kinship between the boy, with his "beaming brow" and "eye and cheek so bright", and the "brilliant" poppy that he "loves to see", with its "gay blossom" and its "petals fresh and fine". Both are united in their freshness and brilliancy: and, just as Herbert brings comfort to the speaker's days, so opium brings comfort to her nights. Unlike her father or De Quincey, however, Sara Coleridge only mentions the relaxant properties of the drug, rather than its potential to stimulate the body or mind. For her, it is simply a way of getting through the night, and if she has dreams under its influence, she does not describe them.

Sara herself drew a distinction between De Quincey's use of opium and her own, writing in a letter of 1825:

> Mr Dequincey is again on sociable terms with the Wordsworth – he is well in tolerable spirits & has left off opium, but Miss W[ordsworth]. now fears he will take to the horrid drug again – *horrid* I call it when thinking of him & some others, in me that is rather ungrateful, as it has done me much good & no harm and I might exclaim with Mrs O Neil "Hail lovely blossom that can'st ease, the wretched victim of disease" […][.] To return to the poppy I could not add with the poetess, "Canst close those weary eyes in sleep that never open but to weep." This happened the summer before last. Night after night has grief & anxiety kept me awake spite of laudanum itself – you will say the dose was not large enough. (cited in Low 2006, 128)

Here, Sara Coleridge distinguishes between De Quincey "& some others" (probably including her father, despite her later attempts to justify his opiate usage in her 1847 edition of *Biographia Literaria*), for whom opium is a "horrid drug" taken for hedonistic reasons, and responsible opium users such as herself, who use it as a sedative that does "much good & no harm" (Coleridge 1847, ii.410). When De Quincey's use of the drug results in negative outcomes, it is because he has not "left off" it entirely; when hers does, it is because "the dose was not large enough". Tellingly, she describes her positive experiences of opium in words borrowed from another female poet: Henrietta O'Neill, whose "Ode to the Poppy" was first published by her friend, Charlotte Smith, in her 1792 novel *Desmond* (Smith 1792, iii.165–66). This choice suggests a gendered dimension to her distinction. Literary men such as De Quincey were suspected of taking opium recreationally, as a stimulant, with destructive consequences for themselves and their families, but Sara Coleridge implies here that for women such as O'Neill and herself it was valuable primarily as a relaxant, providing them with a vital resource in their constant struggle against "grief & anxiety", sleeplessness, sickness and pain.

Sara Coleridge and Henrietta O'Neill were not the only women of the period to publish works about their experiences of opium usage. Other examples include Anna Seward's "Sonnet XIII" (1773) and "To the Poppy" (1810), Mary Robinson's "Ode to Health" (1791), "Invocation to Oberon" (1792) and "Ode to Apathy" (1800), Letitia Landon's "The Poppy" (1838) and Harriet Martineau's *Life in the Sickroom* (1844) and *Letters on Mesmerism* (1845). Jane Carlyle, Mary Russell Mitford, Dorothy Wordsworth, Elizabeth Barrett Browning and George Eliot were also habitual users of opium. For all of these women, opium formed part of ordinary domestic life, used to manage pain, illness and distress, and valued for its power to sedate and tranquilise rather than for its stimulant properties or its ability to induce dreams and visions. The opening lines of Seward's 1773 sonnet—"Thou child of Night and Silence, balmy Sleep / Shed thy soft poppies on my aching brow!"—are typical in this respect: The pain of her "aching brow" is the problem, to which the opium poppy is presented as providing a practical solution. Robinson's "Ode to Apathy", in which "apathy" seems to be a personification of opium reverie, is very similar, with its speaker eagerly looking forward to the moment when "Thy poppy wreath shall bind my brows, / Dead'ning the sense of pain" (Seward 1810, iii.134; Robinson 1824, 58). Robinson's earlier "Ode to Health" includes a passage in which the speaker imagines herself manufacturing such anaesthesia-inducing opiates:

> There I'll press from herbs and flowers
> Juices bless'd with opiate powers,
> Whose magic potency can heal
> The throb of agonizing pain (Robinson 1824, 37)

These poems describe opium being used to manage pain, just as orthodox contemporary physicians recommended, rather than to intoxicate its users. Accordingly, instead of stigmatising opium as dangerous or destructive, they praise it as a "soft" and "bless'd" drug. Among all these texts, only Landon's follows the model of S.T. Coleridge and De Quincey by emphasising the power of opium to induce visions as well as sleep—visions from which its subject is awakened by the kiss of her sister, which breaks the spell of her visionary "slumber" just as the children and the person from Porlock did for De Quincey and S.T. Coleridge, respectively (Landon 1838, n.pag.).

Given the frequency with which opium was used in this period for "female complaints", menstrual pains and the quieting of fretful children, it is not surprising that Robinson, Seward and Sara Coleridge should have depicted it primarily as a form of household medicine rather than as an intoxicant (Berridge 1999, 67–68; 103–04). Insofar as they valued opium for its psychological drug effects, they emphasised its power as a relaxant, and its ability to temporarily banish feelings of worry and depression, rather than its capacity to conjure spectacular hallucinations and dreams. In particular, these poems repeatedly praise the ability of opium to help women cope with feelings of melancholy, especially melancholy induced by past romantic disappointments. In Seward's "Sonnet XIII", it is the loss of "priz'd Affection" which leads the speaker to long for the sleep-inducing poppy: Seward later developed this idea more fully in "To The Poppy", in which the poppy itself is compared to a mad girl standing in a field, its opiate "juice" serving to bring "Kind dreams oblivious" to those who drink it, just as the "brain-sick visions" of the "love-craz'd maid" bring her "false peace" amidst her extreme distress (Seward 1810, iii.134; iii.192). For Seward, while "the votary of love and joy" may prefer roses, "Misfortune's victim" will seek the poppy instead (Seward 1810, iii.192). A very similar idea is expressed by O'Neill in her "Ode to the Poppy", in which the speaker claims that she preferred roses "In early days, when Fancy cheats", but that she has since learned from bitter experience that the soporific poppy "dost so far exceed, / The richest gifts gay Flora can bestow", because it has the power to "disarm" "agonizing Pain" and to "Expel imperious memory from her seat" (Smith 1792, 165–66). In her "Ode to Apathy", Robinson writes:

> When love his tyrant power would prove,
> Thou, vapid dreamer, still to thee
> My darksome pilgrimage shall be (1824, 59)

She here exults in the apathy-inducing power of opium to dull the pains caused by the "tyrant power" of love. These poems thus repeatedly juxtapose girlish figures who favour roses because they still believe in love with older female speakers who have learned to prefer opium poppies because they know that such romantic dreams will only bring them pain and disappointment, reinforcing the association between opium use and older women grappling with the realities and responsibilities of adulthood. That the use of opium to cope with such disappointments could be taken to socially unacceptable extremes is clear from the figure of Lady Delacour in Edgeworth's 1801 novel *Belinda*, whose reliance upon opium to endure her unhappy marriage is presented as highly destructive. In most of these works, however, the sedative power of opiates to banish feelings of misery is depicted as an unalloyed good.

Even Landon's poem, though atypical in other ways, follows this pattern. Describing the tormented opium-dreams of its subject, its speaker declares:

> Mighty must have been the sorrow,

Passionate the grief,
  Which can thus a solace borrow,
From that haunted leaf.

The woman described in the poem has not sought out the "haunted leaf" of the poppy because she is keen to experience the dreams and visions it brings her, but because she is desperate for "solace". The specific nature of the "sorrow" which has afflicted her is hinted at by the very first dream which she experiences when she takes opium:

First, there comes an hour Elysian,
  Would it might remain!
Bringing back Love's early vision,
  But without its pain. (Landon 1838, n.pag.)

As in Seward's poems, it seems to be romantic disappointment which drives Landon's subject, whom she describes as "the broken-hearted", to seek relief in opiates—a disappointment that separates her from her "guileless childhood", during which she, like O'Neill and Seward's speakers, may still have preferred roses to poppies.

I know of only one poem written by a woman during this period which advertised the fact that it was dictated in a state of opium reverie, like that which supposedly inspired S.T. Coleridge to write "Kubla Khan". The work in question is Robinson's "The Maniac", which predated Coleridge's more famous poem by several years: indeed, it has been suggested that Coleridge may have based his story of unconscious, opium-inspired composition of "Kubla Khan" on Robinson's earlier narrative (Hayter 1968, 292; Byrne 2004, 286–87). On the poem's origin, Hayter writes:

> In 1791, when [Robinson] was under treatment at Bath for her rheumatism, which had brought on crippling lameness, she was ordered by her doctor to take 'near eighty drops of laudanum'. Throughout the ensuing night she dreamt about a miserable madman called Jemmy whom she had seen being pelted with stones and mud in the streets of Bath, and whose wretched condition had seized on her imagination. 'Having slept for some hours, she awoke, and calling her daughter, desired her to take a pen and write what she should dictate. Miss Robinson, supposing that a request so unusual might proceed from the delirium excited by the opium, endeavoured in vain to dissuade her mother from the purpose. The spirit

> of inspiration was not to be subdued, and she repeated, throughout, the admirable poem of *The Maniac,* much faster than it could be committed to paper.' [...] On the ensuing morning, Mrs Robinson had only a confused idea of what had passed, nor could be convinced of the fact till the manuscript was produced. (Hayter 1968, 291)

Robinson, however, did not claim to have experienced the "paranoiac visions" of Asian landscapes described by Leask when under the influence of opium (Leask 1992, 171). Instead, she dreamed of the persecuted figure of "Jemmy", whom it is tempting to read as her alter ego. As in so much of the women's opium literature of the period, the poem's focus is not on exaltation or disintegration, but on healing and consolation. After describing the maniac's sufferings, and speculating on what unhappy events may have reduced him to this state, its final stanza reads:

> Oh! tell me, tell me all thy pain;
> Pour to mine ear thy frenzied strain,
> And I will share thy pangs, and soothe thy woes!
> Poor maniac! I will dry thy tears,
> And bathe thy wounds, and calm thy fears,
> And with soft Pity's balm enchant thee to repose. (Robinson 1824, 149)

In many respects, this is a standard sentimental response, reminiscent of the madhouse scene in Henry Mackenzie's famous sentimental novel, *The Man of Feeling* (1771). Given the poem's origin as an opium-vision, however, I find it intriguing that what the speaker imagines herself doing here is exactly what the drug effects of opium are depicted as doing in poems such as "To the Poppy", "Ode to the Poppy" and "Ode to Apathy": soothing woe, calming fear and bringing "repose" to the exhausted sufferer. "Soft Pity's balm" may, here, actually be laudanum, just as the "oblivious dews" that bring "balmy Sleep" to the speaker of Seward's "Sonnet XIII" are almost certainly opiates (Seward 1810, iii.134). Such a reading would be very much in line with the overall depiction of the sedative properties of opium in these poems as a merciful, healing force: Robinson herself certainly relied upon it because of its ability to soothe her physical and emotional pain (Byrne 2004, 285–90). "The Maniac" may indeed have been based upon Robinson's opium-inspired "dream about a madman", but given the severity of her own sufferings by 1791, this image of a tormented individual soothed by "Soft

Pity's balm" could just as easily describe her own relationship with opium as her fantasy encounter with Jemmy in "the streets of Bath".

The distinction I have outlined here, between the stereotype of the visionary male writer who takes opium to escape from quotidian realities and the female writer—often a mother, as Henrietta O'Neill, Sara Coleridge and Mary Robinson all were—who takes opium to help her to cope with those realities and the responsibilities which they bring with them, mirrors the double role played by opium in Romantic-era British culture more generally. If De Quincey's account of his use of opium as a recreational stimulant provoked worries that he was sliding into "oriental" "degeneracy", then Sara Coleridge's depiction of the same drug as a sedative which helped her to manage her duties as a wife and mother allowed her to position her own opium usage within the boundaries of what, at the time, were considered to be the morally and socially acceptable uses of opiates. This same sense of the double nature of opium, as something that was at once strange and threatening and soothing and mundane, is evoked by Robinson's "Invocation to Oberon", in which Oberon declares:

> From the Poppy I have ta'en
> Mortal's balm, and mortal's bane!
> Juice that, creeping through the heart,
> Deadens ev'ry sense of smart;
> Doom'd to heal, or doom'd to kill,
> Fraught with good, or fraught with ill.
> This I stole, when witches fell,
> Busy o'er a murd'rous spell,
> On the dark and barren plain,
> Echo'd back the night-owl's strain! (Robinson 1824, 152)

The "witches fell" presumably values the poppy because it is "doom'd to kill" and "fraught with ill", making it a suitable part of their "murd'rous spell"; but Oberon steals it from them and turns it, instead, into an ingredient of the "healing drops" which restore health to the "aching head" of Robinson's sick daughter, Maria. The opium poppy thus moves from the realm of the exotic, foreign and supernatural (the "dark and barren plain" where the witches brew up their curses) into the realm of domestic reality (Maria's sick-bed), changing from something dangerous into something medicinal in the process. Out *there*, in the world of magic and mysteries evoked by the famous opium-dreams of Coleridge and De

Quincey, opiates might be destructive: but *here*, in the everyday world within which Robinson has to care for her sick daughter, they are simply sedative medicines "fraught with good", an essential part of the domestic arsenal with which Robinson, as a mother, must fight to maintain her own health and the health of her family.

As I have mentioned, the nature of opium dependency was still much debated at the time. It did not escape the notice of doctors such as Reid that some opium users became as dependent on opiates as drunkards were on alcohol. But while S.T. Coleridge and Thomas De Quincey might have been as obviously enslaved by the drug as any Turkish *Theriakis*, their opium habits were primarily viewed as moral rather than medical problems (Reid 1816, 103–06; Zieger 2008, 4–5). Coleridge himself described his addiction as "a species of madness" (Porter and Porter 1988, 224). However, because these women described themselves as using opiates in socially acceptable ways, as medically approved sedatives rather than as recreational stimulants, it was much easier for them to argue that there was nothing pathological about their consumption of opium. Sara Coleridge's distinction between De Quincey's pathological intake and her own healthy, medicinal use of opium has already been mentioned, and a similar resistance to the idea that there was anything pathological about the regular usage of opiates can be seen in Martineau's *Life in the Sick-room* (1844). Martineau described her inability to abandon the use of opium as an example of the way in which invalids tend to become attached to habits and routines rather than as proof of her physical dependence upon the drug:

> The most careless observer must know that it is illness of itself to a sick person to have to wait for food or medicines, or to be put off from regular sleep. Meantime, the invalid cannot keep too careful a watch upon the increase of his own peculiarity – his refuge in custom. There is something shocking to us invalids, when we fix our meditation upon this, in our attachments to our own comforts, and cowardice about dispensing with them. I have myself observed, with inexpressible shame, that, with the newspaper in my hand, no details of the peril of empires, or of the starving miseries of thousands of my countrymen, could keep my eye from the watch before me, or detain my attention one second beyond the time when I might have my opiate. For two years, too, I wished and intended to dispense with my opiate for once, to try how much there was to bear, and how I should bear it; but I never did it, strong as was the shame of always

> yielding; and I have now long given up all thoughts of it. (Martineau 1844, 135–36)

The fact that Martineau could not bring herself to miss a single dose of her opiate for two years, despite her repeated resolutions to do so, is here described as merely another example of the invalid's "refuge in custom", to be classed alongside the discomfort that the sick feel at having their books moved from their proper places. The issue here was not the pain of withdrawal: Martineau's *Life* elsewhere bears witness to her impressive fortitude in the face of physical suffering, and she even wrote about the "cheering" effects of temporary severe pain (Martineau 1844, 113–14). Very clearly she had, by 1844, become dependent upon opiates, but she strongly resisted describing her opium habit as an ailment in its own right, even though it was not until mesmerism provided her with an alternative form of pain relief that she was finally able to break her dependence upon the drug (Martineau 1845, 8–13).

Another example of this tendency would be Elizabeth Barrett's defence of her opiate usage in her letters to Robert Browning. She wrote: "The medical people gave me opium … a preparation of it, called morphine, & ether - & ever since I have been calling it my amreeta [sic] draught, my elixir … because the tranquilizing power has been wonderful". Barrett was clearly surprised by the disapproval that Browning expressed regarding her regular use of morphine and ether—"[T]hat you should care so much about the opium-!" (Browning and Barrett Browning 1990, 210). She explained that even though she did not suffer from chronic pain, she found the drugs essential as sedatives:

> I have had restlessness till it made me almost mad – at one time I lost the power of sleeping quite … and even in the day, the continual aching sense of weakness has been intolerable […] Such a nervous system as I have … so irritable naturally, & so shattered by various causes … that the need has continued in a degree until now - & it would be dangerous to leave off the calming remedy, Mr Jago says, except very slowly & gradually. (Browning and Barrett Browning 1990, 210–11)

Like Sara Coleridge, Anna Seward, Henrietta O'Neill and Mary Robinson before her, Elizabeth Barrett represented opiates (and ether) as valuable and benign, because they were able to soothe the "restlessness" that she found otherwise "intolerable". As Dormandy notes: "To her the drug was a medicine, not an addiction" (Dormandy 2012, 96). Indeed, the symptoms of restlessness and sleeplessness which she describes here are identical to those mentioned by Sara Coleridge in "Poppies". By emphasising the sedative properties of the opiates she used, and insisting that she took them only as relaxants rather than as recreational stimulants, she was able to present her own drug use as something socially appropriate for a wealthy female invalid such as herself, and quite distinct from the hedonistic opium abuse which had come to be associated with male writers such as Coleridge and De Quincey. Indeed, like Martineau, Barrett used opiates precisely because they *helped* her to behave in the ways expected of a female invalid, by quelling the "restlessness" which might otherwise have made her situation unbearable. Like the other female writers I have discussed, Barrett made clear that the sedative properties of opium helped her to carry out the performance of conventional, respectable femininity that was expected of her—and, as such, it was easy for her to insist that her drug usage was benign rather than deviant or anti-social.

Barrett asserted that the primary cause of her "aching sense of weakness" was the mysterious illness which she suffered from, but I find it tempting to wonder whether a secondary cause might not have been the tensions and anxieties inherent in her situation as a female writer working within a cultural climate which viewed the creative and intellectual ambitions of women with great suspicion.[3] Barrett's letters present her regular doses of morphine and ether as helping her to deal with the "restlessness" which kept her from sleep and threatened to drive her "mad"; as helping her, in other words, to cope with the restrictions which encompassed all female writers working in early Victorian Britain, and to function within a household environment dominated by the controlling presence of her overbearing father. Dally, her biographer, notes that she spent almost her entire personal income on morphine and became "restless and upset" when her chemists failed to supply her with sufficiently strong preparations of opium (Dally 1989, 71–72; 79–82; 86–87; 114). It is thus unsurprising that she represented opiates in such a positive light in her

[3] In this context, it seems highly suggestive that even though Erasmus Darwin's biological theories led him to class opium as a "most powerful stimulant", when treating "accomplished" and "ingenious" young women he mostly seems to have used it as a tranquiliser to control their convulsions, spasms, "reveries", headaches and delirium (Darwin 1794, 26–27; 221–23; 364; 427; 439).

letters and resisted Browning's attempts to wean her off the drugs which made her daily life so much easier to bear.

In *Inventing the Addict*, Zieger points out that in the temperance literature of the early nineteenth-century women were almost always portrayed as the long-suffering victims of male selfishness, and that it was not until the 1870s that the popular imagination began to be haunted by the figure of the deceptive female morphine addict (Zieger 2008, 127–28). Before this, habitual opium use was usually not understood in terms of "addiction", and, as a result, these female writers were able to present their opium use as normal and unobjectionable precisely because they located it within the domestic sphere, thus differentiating themselves from the stereotypical male drunkard or "opium-eater" who forsakes his domestic responsibilities in favour of narcotic or alcoholic intoxication. Because opium abuse, like drunkenness, was understood as a moral failing rather than a medical pathology, women who used opiates as sedatives in order to function as better wives, mothers or daughters could hardly be condemned in the same way as opium-eating hedonists who used the same drugs as recreational stimulants. Had the requirements of conventional femininity been less constraining, many of these women would probably never have been driven to take opium in the first place: but when they did, their depiction of opium use as supporting their performance of normative gender roles allowed them to evade censure for using narcotics to help them cope with the demands placed upon them as women, wives and mothers. The resulting works record their gratitude to the opiates which, whatever their disadvantages, helped to protect them from the restlessness, sleeplessness, pain and melancholia which threatened to prevent them from writing or functioning at all.

References

Abrams, M.H. 1934. *The Milk of Paradise: The Effect of Opium Visions on the Works of De Quincey, Crabbe, Francis Thompson, and Coleridge*. Cambridge: Harvard University Press.

Anonymous. 1822. Christen on the Nature of Opium, with the Romance of the English Opium Eater. *The Quarterly Journal of Foreign and British Medicine and Surgery* 4: 350–63. London: John Anderson.

Berridge, Virginia. 1999. *Opium and the People: Opiate Use and Drug Control Policy in Nineteenth and Early Twentieth Century England*. London: Free Association.

Blair, William. 1842. An Opium-Eater in America. *The Knickerbocker: Or New York Monthly Magazine*, July: 47–57. New York: Samuel Hueston.
Brooks, Maria. 1820. *Judith, Esther, and Other Poems*. Boston: Cummings and Hillfard.
Browning, Robert, and Elizabeth Barrett Browning. 1990. *The Courtship Correspondence 1845–1846*, ed. Daniel Karlin. Oxford: Oxford University Press.
Budge, Gavin. 2012. *Romanticism, Medicine, and the Natural Supernatural: Transcendent Vision and Bodily Spectres, 1789–1852*. Basingstoke: Palgrave Macmillan.
Byrne, Paula. 2004. *Perdita: The Life of Mary Robinson*. London: HarperCollins.
Coleridge, S.T. 1847. *Biographia Literaria*, ed. Henry Nelson Coleridge and Sara Coleridge, 2 vols. London: William Pickering.
———. 1969. *Poetical Works*, ed. Ernest Hartley Coleridge. Oxford: Oxford University Press.
Dally, Peter. 1989. *Elizabeth Barrett Browning*. London: Macmillan.
Darwin, Erasmus. 1794. *Zoonomia, or the Laws of Organic Life, volume 1*. London: Joseph Johnson.
De Quincey, Thomas. 1986. *Confessions of an English Opium Eater*, ed. Alethea Hayter. Penguin: London.
Dormandy, Thomas. 2012. *Opium: Reality's Dark Dream*. New Haven: Yale University Press.
Feldman, Paula. 1997. *British Women Poets of the Romantic Era: An Anthology*. Baltimore: Johns Hopkins University Press.
Foxcroft, Louise. 2007. *The Making of Addiction: The "Use and Abuse" of Opium in Nineteenth-Century Britain*. Aldershot: Ashgate.
Grinnell, George. 2006. Thomas Beddoes and the Physiology of Romantic Medicine. *Studies in Romanticism* 45 (2): 223–50.
Hayter, Alethea. 1968. *Opium and the Romantic Imagination*. London: Faber.
Keats, John. 1970. *Complete Poems*, ed. Miriam Allott. London: Longman.
Kondratas, Ramunas. 1988. The Brunonian Influence on the Medical Thought and Practise of Joseph Frank. In *Brunonianism in Britain and Europe*, ed. W.F. Bynum and Roy Porter, 75–88. London: Wellcome Institute.
Landon, Letitia. 1838. The Poppy. In *Flowers of Loveliness*. London: Ackermann and Company.
Lawrence, Christopher. 1988. Cullen, Brown, and the Poverty of Essentialism. In *Brunonianism in Britain and* Europe, ed. W.F. Bynum and Roy Porter, 1–21. London: Wellcome Institute.
Leask, Nigel. 1992. *British Romantic Writers and the East: Anxieties of Empire*. Cambridge: Cambridge University Press.
Low, Dennis. 2006. *The Literary Protégés of the Lake Poets*. Aldershot: Ashgate.
Madden, Richard. 1833. *Travels in Turkey, Egypt, Nubia, and Palestine*. London: Whittaker, Treacher, and Co.

Martineau, Harriet. 1844. *Life in the Sickroom: Essays by an Invalid*. London: Edward Moxon.

———. 1845. *Letters on Mesmerism*. London: Edward Moxon.

Porter, Roy, and Dorothy Porter. 1988. *In Sickness and In Health: The British Experience 1650–1850*. London: Fourth Estate.

Reid, John. 1816. *Essays on Hypochondriacal and Other Nervous Afflictions*. London: Longman, Hurst, Rees, Orme, and Brown.

Risse, Guenter. 1988. Brunonian Therapeutics: New Wine in Old Bottles? In *Brunonianism in Britain and Europe*, ed. W.F. Bynum and Roy Porter, 46–62. London: Wellcome Institute.

Robinson, Mary. 1824. *Poetical Works*. London: Jones and Company.

Seward, Anna. 1810. *Poetical Works*, ed. Walter Scott, 3 vols. Edinburgh: John Ballantyne and Co.

Shiman, Lilian Lewis. 1988. *Crusade Against Drink in Victorian England*. Houndmills: Macmillan.

Smith, Charlotte. 1792. *Desmond*, 3 vols. London: G.G.J. and J. Robinson.

Trotter, Thomas. 1807. *A View of the Nervous Temperament*. London: Longman, Hurst, Rees, and Orme.

———. 1813. *An Essay, Medical, Philosophical, and Chemical, on Drunkenness*. Philadelphia: Anthony Finley.

Vickers, Neil. 1997. Coleridge, Thomas Beddoes and Brunonian Medicine. *European Romantic Review* 8: 47–94.

Young, George. 1753. *A Treatise on Opium*. London: A. Millar.

Youngquist, Paul. 1999. Lyrical Bodies: Wordsworth's Physiological Aesthetics. *European Romantic Review* 10: 152–62.

Wilson, John. 1843. *The Noctes Ambrosianae of "Blackwood"*, 4 vols. Philadelphia: Carey and Hart.

Zieger, Susan. 2008. *Inventing the Addict: Drugs, Race, and Sexuality in Nineteenth-Century British and American Literature*. Amherst: University of Massachusetts Press.

CHAPTER 13

Middlemarch and Medical Practice in the Regency Era: From "Bottles of Stuff" to the Clinical Gaze

Björn Bosserhoff

I have swallowed the weight of an Apothecary in medicine, and what I am better for it, except more patient and less credulous, I know not. —Elizabeth Montagu to the Duchess of Portland (Montagu quoted in Porter [1741] 2001, 32–33)

Ce n'est point une science pour un esprit méthodique; c'est un ensemble informe d'idées inexactes, d'observations souvent puériles, de moyens illusoires, de formules aussi bizarrement conçues que fastidieusement assemblées.—Xavier Bichat on pharmacology in his Anatomie générale (Bichat 1801, xlvi–xlvii)[1]

B. Bosserhoff (✉)
University of Bonn, Bonn, Germany
e-mail: bjoern.bosserhoff@uni-bonn.de

[1] "It is not a science for a methodical and philosophic mind; it is an incongruous combination of erroneous ideas, observations often puerile, means at the best fallacious,

N. Roxburgh and J. S. Henke (eds.), *Psychopharmacology in British Literature and Culture, 1780–1900*, Palgrave Studies in Literature, Science and Medicine,
https://doi.org/10.1007/978-3-030-53598-8_13

In our own twenty-first century, the word *opium* tends to evoke fantasies of the *fin-de-siècle*, of bohemian outlaws, decadent dandies and demimondaines. This is no doubt due to our continued exposure to portrayals of opium consumption in its most "glamourous" form: ever since the late 1800s, literature has championed opium *smoking*, capitalising on De Quinceyian notions of exoticism and blissful immorality. Thus, the opium den—that arch locale of intoxicated indulgence, "replete with the usual suspects, the Lascars and Malays, muttering and mumbling in their drugged isolation" (Foxcroft 2007, 57)—makes famous appearances in stories written by Dickens, Doyle and Wilde; it also features prominently in neo-Victorian texts such as Eleanor Catton's *The Luminaries* (2013). In fact, however, the opium den was rather an anomaly in nineteenth-century Britain, more myth than reality. Instead of being smoked in East End underworld dives, opium was mainly consumed in liquid form and in a decidedly prosaic, domestic context (see Freeman 2012), namely as laudanum, a medicinal tincture made by solving opium in wine, with spices added *ad libitum*.

It is important to understand that virtually all British opium "addicts", including some very illustrious figures, began as *patients*; in comparison, the purely "recreational" consumption of the drug—think De Quincey at the opera—seems neglectable.[2] The typical story of nineteenth-century opium use involves some sort of symptom—from common coughs and colds all the way to "consumption" and "insanity"—for which laudanum was taken. And even though manuals like Buchan's *Domestic Medicine* were wildly popular and self-medication was "for many [...] a dawn chorus" (Porter 1985, 190), most "addiction" cases were iatrogenic in origin, the drug being first prescribed by a doctor. At any rate, laudanum,

and formulae as fantastically conceived as they are preposterously combined" (Bichat 1824, xiv).

[2] Apart from the usual suspects De Quincey and Coleridge, a list of Georgian and Victorian opium "addicts" would include Dr Johnson and his wife, Tetty Porter, Nelson and Wellington, Wilberforce and Mackintosh, Crabbe and Scott, Ada Lovelace and Elizabeth Barrett (before they both turned to morphine instead), Mrs Gaskell and Florence Nightingale, Wilkie Collins and Francis Thompson. To whom may be added George IV himself as well as a choice group of non-British opium consumers such as Benjamin Franklin, Joseph and Jerome Bonaparte, Heinrich Heine, Hector Berlioz, Frederic Chopin, Charles Baudelaire and Mary Todd Lincoln. Since addiction was only "invented" (as a coherent concept) towards the end of the century, it would perhaps be more accurate to term these individuals "habitués" (see Foxcroft 2007, 1–8; Zieger 2008, 1–8).

and thus opium in its then "standard version" (Parssinen 1983, 28), was overwhelmingly (first) taken for medical reasons. What sometimes followed was the sad, old story of tolerance, withdrawal and relapse.[3]

Latter-day semi-synthetic opioids, such as oxy- and hydrocodone, are derivatives of opium alkaloids that were first isolated in the nineteenth century: morphine by Sertürner in 1804, and codeine and thebaine in the 1830s by Robiquet and Pelletier, respectively. Laudanum, based on raw opium, was introduced much earlier—by Thomas Sydenham (1624–89), sometimes designated "The English Hippocrates" or "Dr Opiophile"—and the concoction served as a mainstay of British medicine ever since, celebrated for its sublime analgesic powers.[4] Books and treatises were written on opium's pharmacological effects, best-known among them perhaps John Jones' *Mysteries of Opium Reveal'd* (1700) and Samuel Crumpe's *An Inquiry into the Nature and Properties of Opium* (1793).[5] Slowly, it was worked out that opium was "conveyed with the blood to its sites of action" (Maehle 1995, 70), but by the early nineteenth century there was still considerable controversy as to whether its effects were predominantly sedative or stimulant (see Haller 1989). Among those inclined to take the latter view were adherents to the Brunonian system of medicine: John Brown, an Edinburgh physician, believed opium to be "the most powerful of all the agents that support life, and that restore

[3] A parallel of sorts is the current US opioid crisis, "the worst drug addiction epidemic in US history" according to the Centers for Disease Control (Kolodny 2015). The excessive prescription of opiate painkillers and the subsequent new surge in heroin use have been leading to thousands of overdose deaths every year during the last decade. "Doctors were just throwing pills at her", reports one bereaved widower in Perri Peltz's television documentary *Warning: This Drug May Kill You* (2017). As Peltz's film illustrates, opioid addiction mainly affects white, middle-class Americans—which may very well be why political reactions are so much more "compassionate" (Lopez 2017) than during the 1980s crackdown on crack, which had mainly concerned African Americans.

[4] For the history of laudanum and other traditional opium medicines, see Hodgson (2001, 45–77).

[5] Jones was among the first to warn of (what we would call) opium's horrific withdrawal symptoms: "great, and even intolerable Distresses, Anxieties, and Depressions of Spirits, which in a few days commonly end in a most miserable Death, attended with strange Agonies, unless men return to the Use of Opium" (quoted in Zieger 2008, 117). Similarly, Crumpe's "experiments and observations had shown him that when a regular user was deprived of his or her opium [...] they became 'languid, dejected, and uneasy at the customary hours of taking it, and could only be roused from this state by the usual quantity of Opium, or by a large draught of wine'" (118).

health" (Brown 1795, 242), and throughout the first half of the nineteenth century, his followers would administer great quantities of opium for all sorts of ailments, hoping thus "to raise the degrees of excitement in states of asthenia" (Maehle 1995, 52).[6] Others were more sceptical: In 1832, the toxicologist Robert Christison expressed his astonishment as to "how very little is known by the medical profession of the effects of [...] drinking laudanum on health [and] longevity" (Christison quoted in Foxcroft 2007, 84). Alas, this did not keep doctors from prescribing it, nor people from taking their bottles and jugs to the local druggist to have them filled with "lodlum" on a regular basis: "Everyone had laudanum at home. Twenty or thirty drops could be had for a penny" (Berridge and Edwards 1981, 29).

In what follows, I will revisit "Victorian literature's most famous doctor" (Sparks 2009, 3), George Eliot's Tertius Lydgate, and analyse his stance towards the use of opium in medical practice. In a second step, I will contextualise Lydgate's views within contemporary discourses on pharmacotherapy and medicine in general, arguing that he may be regarded the foremost example in British literature of an early clinician in the Foucauldian mode.

The eminent Victorianist Asa Briggs believed that *Middlemarch* (1871–72) "tells us far more about the significance of medical reform than any other novel of the nineteenth century, and a great deal more than most academic monographs" (Briggs [1948] 1985, 55). Indeed, medicine looms large in Eliot's novel, forming "an integral part" of its "underlying theme of reform and transition" (Furst 1993, 342). As she did for her previous books *The Mill on the Floss* and *Felix Holt, the Radical*—both also set just before the first Reform Act of 1832—Eliot took great pains in researching the late Regency era when working on her "Study of Provincial Life", and at least half of her impressive *Quarry for Middlemarch* is "given over to notes on medical subjects" (Kitchel 1950a, iii). As Patrick McCarthy observes, Eliot and her companion, philosopher George Henry Lewes, were both deeply

> interested in the [medical] profession and had notions for its improvement. Both knew that what had long passed for knowledge among reputedly

[6] It appears that Brown was himself "addicted" to the drug, consuming some 200 drops of laudanum in a glass of whisky during lectures he gave (Brown 1795, lxxxvii; see also Budge 2013, 80; Vickers 1997, 86n23).

learned doctors was "confident ignorance" which appealed to authorities long since discredited and acted on unprovable hypotheses. Lewes's article of 1862, entitled "Physicians and Quacks," said as much and argued for the establishment of a cadre of scientist doctors who could give full time to research [...]. (McCarthy 1970, 812)[7]

Eliot would present her version of a Lewesian "scientist doctor" in Lydgate. As such he clearly stands out among nineteenth-century fictional medical men who for the most part still belong to the tradition of a "long line of quacks and grotesques extending from the mountebanks of Elizabethan drama to Laurence Sterne's Dr. Slop" (McCarthy 1970, 806). Gradually, though, as Q.D. Leavis notes, a more "modern figure", "concerned not for private practice among the well-to-do but for public health and the scientific advancement of medicine" emerged, a figure that became "an evident symbol" of the reformist zeitgeist and, increasingly, "a central character instead of a stock background-filler" (Leavis 1970, 180). Of this type, Lydgate is the prime example. What distinguishes him from forerunners like Harriet Martineau's Edward Hope in *Deerbrook* (1838) or Anthony Trollope's *Doctor Thorne* (1858) is the sheer room Eliot dedicates to descriptions of his medical practice, as well as the decidedly scientific bent she portrays Lydgate's mind to have.[8]

Ever since his early vocational awakening, described in chapter 15 of *Middlemarch*, Lydgate has considered "the medical profession as it might be [...] the finest in the world; presenting the most perfect interchange between science and art; offering the most direct alliance between intellectual conquest and the social good" (Eliot [1871–72] 1994, 145). "[A]s it might be" is the key phrase here: Lydgate is eager to help overcome the deplorable state medicine is in during his own Regency era. Already when the young doctor is first mentioned, we learn that

[7] Eliot and Lewes were themselves afflicted with an "astonishing array of illnesses" (McCarthy 1970, 811) and consequently acquainted with a great number of doctors, including such luminaries as James Paget and William Osler (see Kitchel 1950b, as well as Halstead 1946; Pelling 2006).

[8] In *Middlemarch*, George Levine holds, "[s]cience stands to the text [...] as religion stands to that of *Paradise Lost*" (Levine 1981, 256)—a fact famously singled out by Henry James who complained that Eliot's novel "is too often an echo of Messrs. Darwin and Huxley" (James quoted in Carroll 1971, 359). Little wonder perhaps for an author whose diary comprises remarkable sentences such as this one: "G. and I went to the Museum, and had an interesting morning with Dr. Rolleston who dissected a brain for me" (Levine quoted in Rothfield 1992, 100).

"[m]edical knowledge" is "at a low ebb" (92), and later the novel's narrator characterises the 1820s, supposedly an "age of reform" (French and Wear 1991) for medicine as well, as "a dark period" in which "most medical practice was still strutting or shambling along the old paths" (145; 148). This is what Marshall Hall, MD, just returned from an educational tour on the Continent, had to say about this in his seminal work *On Diagnosis*:

> The conclusions of different inquirers respecting the same point, have been noted as various and contradictory; the evidences on which medical knowledge is founded have been deemed uncertain and equivocal, and the practice of medicine altogether vague and conjectural. (Hall 1817, v)[9]

The fact that medicine was still "largely guesswork" then (Furst 2000, 3) is beautifully illustrated by the "motley assortment" of doctors already established in Middlemarch when Lydgate arrives (Briggs [1948] 1985, 62). We meet the town's two physicians, Drs Sprague and Minchin, who are depicted in caricaturesque contrasts—for example, Sprague "might be expected to grapple with a disease and throw it" whereas Minchin "might be better able to detect it lurking and to circumvent it"—but who "enjoy about equally the mysterious privilege of medical reputation" (182). Then, somewhat lower in the professional pecking-order than the physicians, we have three surgeon-apothecaries: Mr Gambit, who treats his patients to "very full accounts of his own habits of body and other affairs" and whose practice is "much pervaded by the smells of retail trading" (446–47); Mr Toller, cherished for his "active" methods, i.e. for "bleeding and blistering and starving" (447); and Mr Wrench, who is much in demand for his "strengthening" treatments (142), i.e. for supplying his patients with "white parcels" full of drugs (259).[10]

[9] Between Edward Jenner's invention of modern vaccination in 1796 and the first successful demonstration of inhalational anaesthesia fifty years later hardly any progress was made in applied medical research, especially in Britain. This was the time of Babbage's polemical *Reflections on the Decline of Science in England* (1830) and, among many other publications in the same vein, of *An Exposition of the State of the Medical Profession in the British Dominions* (anon 1826), which Eliot read when preparing for *Middlemarch* (see Kitchel 1950b, 6).

[10] See also Eliot's similar depiction of two medical men in "Janet's Repentance": "Pratt elegantly referred all diseases to debility, and, with a proper contempt for symptomatic treatment, went to the root of the matter with port-wine and bark; Pilgrim was persuaded

With their "graspable (and marketable) identit[ies]" and their loyal constituencies, these doctors form a sort of closed system—and the intruder Lydgate is immediately identified as a "threat" to that system's "homeostasis" (Deresiewicz 1998, 731; 733): His rivals becry his professional methods "like a Greek chorus" from the moment he sets foot in their town (Briggs [1948] 1985, 63). But what is it that makes his whole approach to medicine so different from theirs? To begin with, Lydgate's medical education stands in clear contrast to both a physician's humanistic Oxbridge education and the hands-on practices traditionally taught to surgeons. To wit, he attended the prestigious Edinburgh Medical School, followed by a long stay in the mecca of anatomical and clinical instruction in early-nineteenth-century Europe: Paris. As Lilian Furst explains,

> Lydgate's experience of French medical methods is central to his reformist impetus. French clinicians created fresh models of medicine as a science through their ready use of new physical and conceptual tools for examining patients, their analysis of clinical statistics, and their systematic recourse to autopsy. [...] With the end of hostilities after the defeat of Napoleon at Waterloo in 1815, the influx of British medical students to Paris mounted sharply. The name "Gallomaniacs" was coined for British converts to French medical ways on account of their tendency to proselytism. (Furst 1993, 349)[11]

That Lydgate is a member of exactly this "Gallomaniac" cohort of British medical men, Eliot, makes abundantly clear. "I think he is likely to be first-rate – has studied in Paris, knew Broussais; has ideas, you know – wants to raise the profession"—thus Lydgate is lionised by one of Middlemarch's more forward-thinking citizens at the beginning of the novel (92). And Mr Brooke's high hopes for the new doctor are entirely justified: when Lydgate arrives, he is excited by the prospect of doing both "good small work for Middlemarch" and "great work for the world"

that the evil principle in the human system was plethora, and he made war against it with cupping, blistering, and cathartics" (Eliot [1857] 1985, 203). "Heroic" methods for intestinal cleansing would also involve calomel and other highly poisonous forms of mercury.

[11] As regards the long and tangled story of medical professionalisation in the nineteenth century, including the gradual dissolution of the traditional tripartite division of medical men (physician, surgeon, apothecary), M. Jeanne Peterson's 1978 book is still the best resource.

(149). By "careful observation and inference", he hopes to demonstrate "new connections and hitherto hidden facts of structure which must be taken into account in considering the symptoms of maladies and the action of medicaments" (148).[12]

The "action of medicaments" is of particular interest to Lydgate who is keen on making "an advantageous difference to the viscera of his [...] patients" (146). This he intends to achieve by adopting a largely noninterventionist approach with regard to pharmacology: he flatly refuses to dispense medicinal drugs or to use them as the main pillar of his treatments—a decision sure to cause irritation in a town whose inhabitants are used to "swallow[ing] large cubic measures of physic" (146) and who expect to receive "a few bottles of 'stuff'" whenever they consult a doctor (443–44). To make matters worse, they would normally get their medicine *directly* from surgeon-apothecaries like Lydgate (whereas they would take a prescription to the local chemist after consulting a physician). Lydgate's innovation to charge his patients not for dispensing drugs but for his professional attendance—in accordance with "a recent legal decision" (147)[13]—thus proves equally "offensive" to his rival doctors and "to the laity" (444). Eliot here illustrates that around 1830 general practitioners were still "a hybrid class in an anomalous position" (Furst 1998, 66): they were likely to be seen as "arrogating the privilege of a physician" while their dual qualification as surgeon-apothecary, which rapidly became the norm after the 1815 Apothecaries Act, to many "still smack[ed] of trade on account of the use of hands and the dispensing of drugs" (Furst 1993, 354; 343).

As Lydgate explains to the grocer Mr Mawmsey, in not selling drugs he is very much motivated by deliberations concerning professional honour:

[12] In true Romantic fashion, Lydgate dreams of discovering "an organising principle [...] which would bind together disparate parts and reveal unity beneath apparent chaos" (Shuttleworth 1984, 146). It has often been pointed out that, in this respect, he resembles another *Middlemarch* character, Mr Casaubon, who (just as futilely) pursues the "Key to all Mythologies" (Eliot [1871–72] 1994, 63). In reality, the "irreducible structural and active element of vital organization" (Coleman [1971] 1999, 21) was identified in the late 1830s by Schleiden and Schwann: the cell. Bacteriology followed suit, and thus a revolution in disease control.

[13] Probably that of the Handey v. Henson case of 1829, discussed at length in *The Lancet* that year; see Katz (1998). In many of his opinions regarding medical professionalism, Lydgate clearly echoes *The Lancet*'s quarrelsome first editor, Thomas Wakley (cf. Eliot [1871–72] 1994, 156–57; Eliot 1950 *passim*).

"[I]t must lower the character of practitioners [...] if their only mode of getting paid for their work was by their making out long bills for draughts, boluses, and mixtures" (444). Mawmsey, a staunch believer in "the massive benefit of [...] drugs to self and family", is taken aback by this:

> For years he had been paying bills with strictly made items, so that for every half-crown and eighteen-pence he was certain something measurable had been delivered. He had done this with satisfaction, including it among his responsibilities as a husband and father, and regarding a longer bill than usual as a dignity worth mentioning. (445)

The grocer's wife is even more indignant:

> Does this Mr. Lydgate mean to say there is no use in taking medicine? [...] I like him to tell me how I could bear up at Fair time, if I didn't take strengthening medicine for a month beforehand. [...] Does he suppose that people will pay him only to come and sit with them and go away again? (446)

Henceforth, Lydgate is preceded by a dubious reputation: "[I]t was inevitable that those who called him in should watch a little anxiously to see whether he did 'use all the means he might use' in the case" (449). "[Y]ou should send out medicines as the others do", even his wife Rosamond advises him (649), but to no avail.

A brief look at the actual cases treated by Lydgate reveals that his doctoring indeed mainly consists of "sitting with" people. His "first serious transgression" of Middlemarch medical conventions (Greenberg 1975) occurs in chapter 26 when he happens to walk by the Vincys' house and is asked inside, regardless of all "medical etiquette", by the mayor's wife who is worried for the health of her son. Given Fred Vincy's symptoms—"depression and headache", followed by pronounced malaise and lots of "shivering by the fire"—their family doctor, Mr Wrench, had diagnosed merely a "slight derangement" but nonetheless sent his "usual white parcels which this time had black and drastic contents". If anything, these appear to make Fred's condition worse, and Lydgate, "convinced that Fred was in the pink-skinned stage of typhoid fever, and that he had taken just the wrong medicines", prescribes bed rest and intensive attendance, with only minimal medicinal aid. After a "critical stage" of "delirium", Fred ultimately recovers, and the Vincys gladly spread the tidings that Lydgate is "wonderfully clever in fevers".[14] But the episode does not go down well with the embarrassed Mr Wrench, who accuses Lydgate of using "tricks, worthy only of a quack" and decries his "flighty, foreign notions" and "ungentlemanly attempts to discredit the sale of drugs by his professional brethren" (259–65).

Throughout the first half of *Middlemarch*, we are treated to variations of the Vincy case. When attending to Mr Casaubon, Lydgate basically sits "quietly by his patient and watche[s] him", eventually prescribing "moderate work" and "variety of relaxation" (286). In the case of Mr Trumbull, "an eloquent auctioneer [...] seized with pneumonia", he likewise proceeds by "watching the course of [the] disease when left as much as possible to itself" (451). Then, there is "the speedy restoration of Nancy Nash" (451), a charwoman who is

> afflicted with a tumour at first declared to be as large and hard as a duck's egg, but later in the day to be about the size of 'your fist'. Most hearers agreed that it would have to be cut out, but one had known of oil and another of 'squitchineal' as adequate to soften and reduce any lump in the body when taken enough of into the inside – the oil by gradually 'soopling', the squitchineal by eating away. (450)

This time contradicting Dr Minchin's previous diagnosis, Lydgate declares that Nancy suffers merely from abdominal cramps: "He ordered her a blister and some steel mixture, and told her to go home and rest, giving her at the same time a note [...] to testify that she was in need of good food" (450).

[14] The "nature of fever or fevers" is one of Lydgate's special interests (Eliot [1871–72] 1994, 147): the new hospital that he and Bulstrode are planning to build, envisioned as "the nucleus of a medical school" (124), is meant to be dedicated to "fever in all its forms; Lydgate was to be chief medical superintendent, that he might have free authority to pursue all comparative investigations which his studies, particularly in Paris, had shown him the importance of ..." (453). In this focus on fevers, he is a (fictional) successor to Broussais (see Foucault [1963] 1973, 174–99) and an ally of British sanitary reformers like Edwin Chadwick and Thomas Southwood Smith, whose efforts would eventually result in the 1848 Public Health Act (see Pelling 2006, 227–28). The people of Middlemarch, however, react to Lydgate's hospital plans with considerable fear, believing that what he really wants to do is "try experiments" on them and "kill a few people for charity" (Eliot [1871–72] 1994, 92). These were the days, after all, of the infamous Burke and Hare who murdered sixteen people and sold the bodies to an Edinburgh anatomist for dissection purposes (see Richardson 1987, 132–47).

Finally, we need to consider the case of Raffles, an alcoholic suffering from a life-threatening attack of delirium tremens.[15] When introduced to Raffles, Lydgate "still think[s] he may come round", and he directs his associate Mr Bulstrode to administer "extremely moderate doses of opium" to him (706).[16] What he does not know is that Raffles is a "haunting ghost" of Bulstrode's past (717), that he has come to Middlemarch to blackmail Bulstrode and that, consequently, the latter has no interest whatsoever in seeing him recover. Thus, Lydgate's instructions to cease the opium doses at some point and under no circumstances to give the patient any additional alcohol are being knowingly ignored—and Raffles dies, an event that contributes to Lydgate's eventual downfall. For our purposes, what matters here are Lydgate's yet again rather conservative therapeutic measures when faced with Raffles' "withdrawal symptoms", compared to both what was usual then (more alcohol) and what is now (benzodiazepines). Being "strongly convinced against the prevalent practice of allowing alcohol and persistently administering large doses of opium" (as, for instance, Brunonians would do) and having "repeatedly acted on this conviction with a favourable result" (700), he only fails in this case because of Bulstrode's personal vendetta against the patient in question.[17]

[15] Eliot's narratives of (what we would now term) addiction are for the most part "stories of ruin" (McCormack 2000, 56): one thinks of Molly Farren in *Silas Marner* who tragically overdoses on laudanum, or of various "pathetic minor characters" who, like Raffles, "drink themselves into oblivion or death" (Warhol 2002, 100–01). In *Middlemarch*, rather more harmless habits prevail; thus, Lady Chettam is "addicted" to "homemade bitters" (90) and Fred Vincy to gambling (Eliot [1871–72] 1994, 670–72). Then again, when Bulstrode says that his "imperfect health has induced [him] to give some attention to those palliative resources which the divine mercy has placed within our reach" (125) he is almost certainly referring to opium. Also, among the "experiment[s] in [...] ecstasy" Will Ladislaw has undertaken is one that involves "ma[king] himself ill with doses of opium"—though "[n]othing greatly original had resulted from these measures; and the effects of the opium had convinced him that there was an entire dissimilarity between his constitution and De Quincey's" (83). Similarly, Lydgate himself "had once or twice tried a dose of opium" but realised that "he had no hereditary constitutional craving after such transient escapes from the hauntings of misery" (668).

[16] Later on, a "phial" is mentioned (711), so clearly Lydgate's medicine of choice is laudanum.

[17] Lydgate exercises the caution suggested by Thomas Trotter in his *Essay on Drunkenness* (1804, 34–35) and appears to follow more or less precisely the instructions given in Thomas Sutton's *Tracts on Delirium Tremens* (1813, 15; 27; 35). Sutton dismisses

At the heart of Lydgate's mission, then, and of *Middlemarch*'s narrative of medical reform, is the question of pharmaceuticals, whose widespread use he intends to curb (notwithstanding his own occasional recourse to "steel mixture" or "moderate doses of opium"; see above). Lydgate's stance on these matters echoes several pioneers of late Regency medicine. For one thing, Eliot was inspired by the views of Edinburgh physician John Thomson, who argued in his *Account of the Life, Lectures, and Writings of William Cullen* (1832):

> The doubts that have till within a short period been entertained with regard to the legal rights of General medical practitioners in England, to claim remuneration for their professional attendance and advice, and the encouragement thus held out to them, to make their professional emoluments depend on the quantity of useless, and too often pernicious, drugs with which they may be able to drench their patients, have been equally prejudicial to the interests of the Public, and degrading to the Medical profession; and, so long as the English public shall chuse to foster this most absurd and disgraceful trade in drugs, now very generally abandoned in this part of the kingdom [i.e. Scotland], it is in vain to expect that medicine can be practised amongst them, upon scientific, rational, or honourable principles. (quoted in Pratt and Neufeldt 1979, xxxvii)[18]

However, Lydgate's anti-drug rationale involves certain medical-scientific convictions, adopted while studying in Edinburgh and Paris, just as much as questions of professional honour. What he goes up against is the tendency among the old guard of doctors to resort to shotgun polypharmacy, vaguely hoping to vitalise their patients' entire bodily systems all at once. As we have seen, the "'strengtheners and the lowerers' still held sway" in Lydgate's early 1830s, "and medicines were still being designated according to whether they were deobstruent or incisive; diluent or attenuant; inviscant or incrassant; astringent or relaxant; refrigerant or calorific" (Halstead 1946, 419). But traditional Galenic medicine

the classic Brunonian remedy for just about any symptom—great quantities of opium combined with spirits—as "a recommendation to a fit of inebriety" (217).

[18] Always resonating in such criticisms is "suspicion of the medical profession itself – that it exacerbated existing ailments, or even caused new ones, in order to promote its own business" (Zieger 2008, 133), in other words: of the dangers of medicalisation (see Illich 1975).

with its emphasis on balancing bodily fluids as well as late-eighteenth-century reformist schools like John Brown's, which relied on opium as an all-important stimulant, would increasingly come under attack. Thus, Andrew Combe, another Edinburgher admired by Eliot (see McCormack 2000, 167), joined Thomson in complaining that "many medical men [...] considered their only business to be the prescription of drugs" (A. Combe 1846, 506). Instead, Combe thought doctors ought to "read aright the indications of Nature" (507) and

> exercise, throughout the whole course of the disease, the most watchful observation over its phenomena and progress [...]. In this way the physician may often exercise the most salutary influence, nay, even be the means of saving the patient's life, and yet not give one particle of medicine. [...]
>
> So far from being always necessary to a cure, drugs are required only where the power of Nature to resume her normal action proves inadequate [...]. (508, 520)[19]

To be sure, Combe makes for a peculiar read: In his *Vis medicatrix naturae* conception, "Nature" is but a synonym for the Christian God, and he was sympathetic to the views of his older brother George, the chief British proponent of phrenology. And yet, like Lydgate, Combe was a "Gallomaniac": He had studied in Paris in 1817–18, where he was mentored by the esteemed anatomist Dupuytren at the Hôtel-Dieu (see G. Combe 1850, 53–54).

In "The Patient's View: Doing Medical History from Below", Roy Porter influentially criticised medical historians for clinging to "the old Whiggish mainline from magic to medicine"—and thus failing to account for the hosts of "fringe" practitioners (like herbalists or hydrotherapists) who still constituted the de facto majority among medical men in the early and mid-nineteenth century (Porter 1985, 194). Old and new medical theories and practices co-existed somewhat uneasily then, and proponents of the Paris school and of "alternative medicine" alike confronted the "the hocus-pocus of the regulars" and accused mainstream doctors of "erroneously striving to destroy disease with poisonous drugs" (Porter

[19] In his bestselling *Treatise on the Physiological and Moral Management of Infancy*, Combe likewise warned nurses of "[d]efects to be guarded against – intemperance the worst – mischievous tendency to administer medicine on all occasions" (Combe 1840, 219), and mothers of "the habit of resorting immediately to laudanum, sedative drops, poppy syrup, spirits, and other means of forcing sleep" on their children (286–87).

1989, 231; 2001, 208).[20] Of all "fringe" movements, the most persistent of course was homeopathy, a system virtually defined by its rejection of (conventional) pharmaceutical drugs. What its founder, Samuel Hahnemann, wrote in the preface to the fifth edition of his *Organon* will sound rather familiar by now:

> the old school of medicine [...] assails the body with large doses of powerful medicines, often repeated in rapid succession for a long time, whose long-enduring, not unfrequently frightful effects it knows not, and which it, purposely it would almost seem, makes unrecognisable by the commingling of several such unknown substances in one prescription, and by their long-continued employment, it develops in the body new and often ineradicable medicinal diseases. (Hahnemann [1810] 1849, vii–viii)

Lydgate's doctrines are also clearly reminiscent of Louis Vitet's *méthode expectante*, as influentially advocated by René Laënnec (and ultimately going back to Hippocrates). Around 1800, there had been "a revival of the old dispute between active medicine and expectant medicine" (Foucault [1963] 1973, 17), and Vitet, a former mayor of Revolutionary Lyon, was one of the main proponents of the latter approach. His doctrine proved far more popular in his native France than in Britain where his principal work *Médecine expectante* was universally panned in the popular reviews—even though one anonymous reviewer grudgingly conceded that oftentimes more may be accomplished by "quietism, than the most vigorous assailants have obtained by their hardihood" (*New Annual Register* 1804, 336). Another representative British opinion is that of fashionable London physician Charles Scudamore who in 1826 expressed his misgivings regarding "the favourite method of practice with many of the French physicians": "[T]o look on passively as a spectator, and witness the formation of disease, forbearing all active treatment [...] seems more reprehensible even than excess of zeal in the use of medicine"

[20] An exemplary figure in whom these tendencies met was François Broussais, Lydgate's teacher at the Val-de-Grâce, who swore by bloodletting via leeches but was fervently opposed to purging and the consumption of stimulants (see Bynum 1994, 45–46). For the "medical fringe" up to mid-century, see also Bynum and Porter (1987). In the context of anti-medical movements, one also thinks of Christian Science, founded in the 1870s by (lifelong secret morphine addict) Mary Baker Eddy.

(Scudamore 1826, 62).[21] The French medical journalist Louis Peisse, by contrast, praised Vitet's revival of a Hippocratic "medicine of observation", arguing that an expectant stance was certainly indicated "when there were no good reasons for more energetic procedures"; these he condemned in unison as being "all more or less murderous" (Peisse quoted in La Berge 2002, 198).

As *Middlemarch* shows, while around 1830 most practitioners in Anglophone countries still clung to age-old vitalist medical beliefs and busily engaged in "heroic" treatments, those who had been educated in France favoured a *primum non nocere* view ("first, do no harm"). This is also illustrated by the case of Oliver Wendell Holmes, American physician-poet, who walked the wards of the Hôpital de la Pitié with Pierre-Charles-Alexandre Louis in the mid-1830s and on returning to the United States would help make the "expectant method" a "hallmark of Boston medicine" (Dowling 2006, 71). "A partial answer" to the customary helplessness of doctors, Holmes' biographer explains,

> would be found by Paris-trained physicians in the *méthode expectante*, the practice of doing everything possible to aid nature in the processes of recovery from disease, and to do nothing to hinder what we have heard Laennec [...] call *la nature médicatrice* – no small accomplishment in the day when physicians, backed by a thousand years of Galenic or humoral medicine, were trained to attack disease vigorously with bleeding, emetics, purgatives, blisters, and a myriad other ostensible remedies meant, as the American physician Benjamin Rush would put it, to drive Nature [...] out the door. (Dowling 2006, 43–44)

Not least, what students were taught in the Paris hospitals—by men like Louis, whose "anatomical demonstrations" Lydgate is said to have attended (164)—is what Foucault in *Naissance de la clinique* famously referred to as the "positive gaze" of the clinician. That gaze (*le regard*), Foucault writes, "has the power to bring a truth to light" ([1963] 1973, xiii), a truth "residen[t] [...] in the dark centre of things" (xiii), in the "silent world of the entrails" (xi) hitherto all but inaccessible to sensory

[21] Generally, British and American observers tended to "denounce and shun" French therapeutic practice in particular: "French clinicians were stereotyped as ardent in their desire to understand disease but apathetic in their efforts to cure" (Warner [1998] 2003, 251).

perception. In a literal sense, this was made possible by major improvements in the fields of pathological anatomy and physiology, especially by Xavier Bichat, a hero of Lydgate's along with Andreas Vesalius. When it came to *living* patients, *le regard* enabled the doctor to "emulat[e]" these very "anatomical dissections that inaugurated the clinical era" (Kennedy 2010, 9). It is "a gaze of the concrete sensibility" (Foucault [1963] 1973, 120), "pure" and "faithful to the immediate" (107), "unprejudiced" and "free [...] of theories and chimeras" (195)—for, as Foucault quotes another icon of Paris medicine, Jean-Nicolas Corvisart, "theory falls silent or almost always vanishes at the patient's bedside to be replaced by observation or experience" (120). What is more, the observing gaze is entirely "gestureless": The clinician will "listen" to the disease but mostly "refrain[...] from intervening", largely "leav[ing] things as they are" (107). Pure Lydgate, yet again.[22]

By championing "hard evidence", retrieved by means of thorough examination, up against the often anecdotal disease interpretations of patients themselves, clinical medicine empowered doctors. And in the quest for signals from the *milieu intérieur* (Claude Bernard), they were soon able to resort to new instruments: the stethoscope was introduced by Laënnec in 1816 (and used by Lydgate some fifteen years later; see Eliot [1871–72] 1994, 286), to be "eventually followed by an increasingly invasive set of 'scopes' (the laryngoscope, the bronchoscope, the endoscope, and so forth)" (Furst 1998, 61)—instruments, that is, which served as extensions of the doctor's senses in identifying abnormalities. As a result, medical practice was "no longer a two-dimensional reading of symptoms, but a three-dimensional probing from symptomatic surface into diseased interior" (Long 1992, 137), a process further spurred by later diagnostic techniques like radiography, sonography or tomography. Before most of these devices were invented, though, doctors had to rely on their "gaze". Lydgate is very much an early-nineteenth-century clinician in the Foucauldian sense, a master in "the art of observing" (Foucault [1963] 1973, 108). Whenever we witness him treating patients, his

[22] Foucault was of course aware of the quasi-religious undertones of his deliberations, which appear to claim almost a *vates* role for the physician. In fact, while not denying its actual power, he saw the concept of the clinical gaze as a self-serving feature of the medical profession's attempts to enhance its own status and capability of exerting social control (see Foucault [1963] 1973, 88–89). For the survival of notions of Romantic medicine in a time dominated by scientific positivism (in *Middlemarch* and beyond), see Tambling (1990).

demeanour is most diligent and correct, displaying “detached concern”. Again and again, he proves that he is well qualified to read and interpret bodily signs, quickly grasp the aetiology of a disease and identify the most promising therapeutic measures. He seems aware that true medical insights can “only be gained by cultivating [a] distanced view, one that paradoxically also sees more closely” (Kennedy 2010, 9).

“The strongest single impulse” of French medicine, John Harley Warner finds, “was directed toward moderation” in terms of medication: “from drugging to knowing” (Warner [1986] 1997, 185; 267). His Paris training gives Lydgate an advantage over his rivals; it enables him to see individuals instead of merely symptoms and also clearly affects his pharmaceutical (non-)practices: long before F. E. Anstie’s *Stimulants and Narcotics* (1864), which “provided a definitive outline of the actions of drugs, including any addictive properties they might have” (Foxcroft 2007, 142), Lydgate wisely exercises caution. It is exactly this “new kind of doctor” (Harvey 1967, 28) that Lewes endorsed when he distinguished between “physicians and quacks” in his eponymous article:

> It would be easy to collect a small volume of telling citations to prove the preposterous opinions which have at various times determined the practice of medicine [...]. But there has been progress [...]. New lights from science have illuminated what was before obscure. And by the dint of the perpetual insurgency of the sceptical spirit, the art and practice of Medicine have slowly advanced. (Lewes 1862, 167–68)

Medical skill, Lewes believes,

> mainly consist[s ...] in accurately determining “what is the matter with the patient.” This may seem an easy thing. It is the main difficulty of the art. It is the guide of practice. [...] The wise physician knows that the whole mystery of Medicine lies just here—in correctly discerning what *are* the indications of a particular malady, and in correctly discriminating what are the effects of certain remedies. [...]
>
> the Physician and the Quack [...] both are ignorant of the real order and succession of the phenomena arising out of the administration of the drug. But the confidence of the Physician is relative and tentative; the confidence of the Quack is absolute and final. [...] The Quack never doubts, never watches. (169–70)

According to this mid-Victorian definition, then, the surgeon-apothecary Lydgate is a true "physician", while his disease- and drug-mongering university-educated rivals are really just quacks. Alas, their patients, Eliot's provincial late-1820s Midlanders, do not care. Like William Lawrence and other Regency-era scientific pioneers, Lydgate bears the brunt of ever-vigilant British anti-Gallicanism, hostile to anything that smacks of "materialism" or, God forbid, "atheism" (see Holmes 2009, 318): He "ought to have kept among the French", the Middlemarch gossips come to agree (744). If the French Revolution in politics did not make it across the Channel, neither did that in medicine, at least not right away. And perhaps this should not even surprise us. It is easy from hindsight to ridicule pre-modern medicine and underestimate the "psychic gratifications offered" by that system, "fraudulent though it may be" (Deresiewicz 1998, 729). As J. Russell Reynolds, future president of the Royal College of Physicians, pointed out in his address before the North London Medical Society in 1858, the majority of his countrymen still fell into the category of those

> who believe in the orthodox practitioners; who have faith in the Royal College of Physicians and Surgeons, and in the Worshipful Society of Apothecaries; who call things by old names [...]; respect fervently leeches, blisters, black draughts, blue pills, and colocynth; [... . T]hey respect most highly some one who "knows their constitution," and who knew their father's, [...] and, giving themselves up to the man who brought them into the world, or sent their fathers out, they are bothered with no doubts, and are content to be killed or cured in the approved and orthodox old way. (Reynolds 1858, 19)

And thus, at the end of the novel, we leave Middlemarch just as it was before Lydgate's arrival, with men like Minchin and Wrench still happily "kill[ing] or cur[ing]" as they always have. Our hero, however, becomes an anti-hero in Eliot's "Finale": Lydgate dies at age fifty in "a Continental bathing-place" (834)—resigned, disillusioned, "regarding himself as a failure" (835).

References

Berridge, Virginia, and Griffith Edwards. 1981. *Opium and the People: Opiate Use in Nineteenth-Century England*. London: Allan Lane.

Bichat, Xavier. 1801. *Anatomie générale, appliquée à la physiologie et à la medicine*. Paris: Brosson, Gabon et Cie. https://wellcomelibrary.org/item/b24923746. Accessed 7 February 2020.

———. 1824. *General Anatomy, Applied to Physiology and to the Practice of Medicine*, trans. Constant Coffyn and George Calvert. London: Highley. https://wellcomelibrary.org/item/b29340445. Accessed 7 February 2020.

Briggs, Asa. 1985 [1948]. *Middlemarch* and the Doctors. In *The Collected Essays of Asa Briggs. Volume II: Images, Problems, Standpoints, Forecasts*, 49–67. Brighton: Harvester Press.

Brown, John. 1795. *The Elements of Medicine* 1. London: Johnson. https://books.google.de/books?id=x2M_AAAAcAAJ. Accessed 7 February 2020.

Budge, Gavin. 2013. *Romanticism, Medicine and the Natural Supernatural: Transcendent Vision and Bodily Spectres, 1789–1852*. Basingstoke: Palgrave Macmillan.

Bynum, W.F. 1994. *Science and the Practice of Medicine in the Nineteenth Century*. Cambridge: Cambridge University Press.

Bynum, W.F., and Roy Porter (eds.). 1987. *Medical Fringe & Medical Orthodoxy, 1750–1850*. London: Croom Helm.

Carroll, David (ed.). 1971. *George Eliot: The Critical Heritage*. London: Routledge & Kegan Paul.

Coleman, William. 1999 [1971]. *Biology in the Nineteenth Century: Problems of Form, Function and Transformation*. Cambridge: Cambridge University Press.

Combe, Andrew. 1840. *A Treatise on the Physiological and Moral Management of Infancy*, 4th ed. Edinburgh: Machlachlan, Stewart, & Co. https://wellcomelibrary.org/item/b21987890. Accessed 7 February 2020.

———. 1846. On the Observation of Nature in the Treatment of Disease. *The British and Foreign Medical Review* 41: 505–24. https://books.google.de/books?id=ORoUAAAAQAAJ. Accessed 7 February 2020.

Combe, George. 1850. *The Life and Correspondence of Andrew Combe, M.D.* Edinburgh: Machlachlan and Stewart. https://archive.org/details/lifecorrespondenOOcombiala. Accessed 7 February 2020.

Deresiewicz, William. 1998. Heroism and Organicism in the Case of Lydgate. *Studies in English Literature, 1500–1900* 38: 723–40.

Dowling, William C. 2006. *Oliver Wendell Holmes in Paris: Medicine, Theology, and The Autocrat of the Breakfast Table*. Durham: University of New Hampshire Press.

Eliot, George. 1950. *Quarry for Middlemarch*, ed. Anna T. Kitchel. Berkeley: University of California Press.

———. 1985 [1857]. Janet's Repentance. In *Scenes of Clerical Life*, ed. George Eliot, 189–334. Oxford: Clarendon Press.

———. 1994 [1871–72]. *Middlemarch: A Study of Provincial Life*. London: Penguin.

———. 2005 [1861]. *Silas Marner*. New York: Simon & Schuster.
Foucault, Michel. 1973 [1963]. *The Birth of the Clinic: An Archaeology of Medical Perception*, trans. A.M. Sheridan Smith. London: Tavistock.
Foxcroft, Louise. 2007. *The Making of Addiction: The "Use and Abuse" of Opium in Nineteenth-Century Britain*. Aldershot: Ashgate.
Freeman, Hannah Cowles. 2012. Opium Use and Romantic Women's Poetry. *South Central Review* 29: 1–20.
French, Roger, and Andrew Wear (eds.). 1991. *British Medicine in an Age of Reform*. London: Routledge.
Furst, Lilian R. 1993. Struggling for Medical Reform in *Middlemarch*. *Nineteenth-Century Literature* 48: 341–61.
———. 1998. *Between Doctors and Patients: The Changing Balance of Power*. Charlottesville: University Press of Virginia.
———. 2000. *Medical Progress and Social Reality: A Reader in Nineteenth-Century Medicine and Literature*. Albany: State University of New York Press.
Greenberg, Robert A. 1975. Plexuses and Ganglia: Scientific Allusions in *Middlemarch*. *Nineteenth-Century Fiction* 30: 33–52.
Hahnemann, Samuel. [1810] 1849. *Organon of Medicine*, trans. R.E. Dudgeon. London: Headland. https://wellcomelibrary.org/item/b21307957. Accessed 7 February 2020.
Hall, Marshall. 1817. *On Diagnosis*. London: Longman. https://wellcomelibrary.org/item/b21271793. Accessed 7 February 2020.
Haller, John S. 1989. Opium Usage in Nineteenth Century Therapeutics. *Bulletin of the New York Academy of Medicine* 65: 591–607.
Halstead, Frank G. 1946. George Eliot: Medical Digressions in *Middlemarch*, and Eliot's State of Health. *Bulletin of the History of Medicine* 20: 413–25.
Harvey, W.J. 1967. The Intellectual Background of the Novel: Casaubon and Lydgate. In *Middlemarch: Critical Approaches to the Novel*, ed. Barbara Hardy, 25–37. London: Athlone Press.
Hodgson, Barbara. 2001. *In the Arms of Morpheus: The Tragic History of Morphine, Laudanum and Opium and Patent Medicines*. Buffalo: Firefly Books.
Holmes, Richard. 2009. *The Age of Wonder: How the Romantic Generation Discovered the Beauty and Terror of Science*. London: Harper Press.
Illich, Ivan. 1975. *Medical Nemesis: The Expropriation of Health*. London: Calder & Boyars.
Katz, Leslie. 1998. An End to Converting Patients' Stomachs into Drug-Shops: Lydgate's New Method of Charging His Patients in *Middlemarch*. *George Eliot—George Henry Lewes Studies* 34/35: 48–59.
Kennedy, Meegan. 2010. *Revising the Clinic: Vision and Representation in Victorian Medical Narrative and the Novel*. Columbus: Ohio University Press.

Kitchel, Anna Theresa. 1950a. Preface. In *Quarry for Middlemarch*, by George Eliot, iii–v. Berkeley: University of California Press.

———. 1950b. Introduction. In *Quarry for Middlemarch*, by George Eliot, 1–19. Berkeley: University of California Press.

Kolodny, Andrew. 2015. The Prescription Opioid Epidemic and the Heroin Revival. Paper presented at the *Addictions Old and New* Conference. University of Richmond, October 22. https://history.richmond.edu/addiction-conference. Accessed 7 February 2020.

La Berge, Ann. 2002. The Rhetoric of Hippocrates at the Paris School. In *Reinventing Hippocrates*, ed. David Cantor, 178–99. Aldershot: Ashgate.

Leavis, Q.D. 1970. Appendix A: The Symbolic Function of the Doctor in Victorian Novels. In *Dickens the Novelist*, by F.R. Leavis and Q.D. Leavis, 179–83. London: Chatto & Windus.

Levine, George. 1981. *The Realistic Imagination: English Fiction from Frankenstein to Lady Chatterley*. Chicago: University of Chicago Press.

Lewes, George Henry. 1862. Physicians and Quacks. *Blackwood's Edinburgh Magazine* 91: 165–78. https://archive.org/details/blackwoodsmagazi91edinuoft. Accessed 7 February 2020.

Long, John C. 1992. Foucault's Clinic. *Journal of Medical Humanities* 13: 119–38.

Lopez, German. 2017. When a Drug Epidemic's Victims Are White. *Vox*, April 4. https://www.vox.com/identities/2017/4/4/15098746/opioid-heroin-epidemic-race. Accessed 7 February 2020.

Maehle, Andreas-Holger. 1995. Pharmacological Experimentation with Opium in the Eighteenth Century. In *Drugs and Narcotics in History*, ed. Roy Porter and Mikuláš Teich, 52–76. Cambridge: Cambridge University Press.

Martineau, Harriet. 1983 [1839]. *Deerbrook*. London: Virago Press.

McCarthy, Patrick J. 1970. Lydgate, "The New, Young Surgeon" of *Middlemarch*. *Studies in English Literature, 1500–1900* 10: 805–16.

McCormack, Kathleen. 2000. *George Eliot and Intoxication: Dangerous Drugs for the Condition of England*. Basingstoke: Palgrave Macmillan.

Parssinen, Terry M. 1983. *Secret Passions, Secret Remedies: Narcotic Drugs in British Society, 1820–1930*. Philadelphia: Institute for the Study of Human Values.

Pelling, Margaret. 2006. Scenes from Professional Life: Medicine, Moral Conduct, and Interconnectedness in *Middlemarch*. In *Politics and Culture in Victorian Britain: Essays in Memory of Colin Matthew*, ed. Peter Ghosh and Lawrence Goldman, 220–36. Oxford: Oxford University Press.

Peterson, M. Jeanne. 1978. *The Medical Profession in Mid-Victorian London*. Berkeley: University of California Press.

Porter, Roy. 1985. The Patient's View: Doing Medical History from Below. *Theory and Society* 14: 175–98.

———. 1989. *Health for Sale: Quackery in England, 1660–1850*. Manchester: Manchester University Press.

———. 2001. *Bodies Politic: Disease, Death and Doctors in Britain, 1650–1900*. Ithaca: Cornell University Press.

Pratt, John Clark, and Victor A. Neufeldt. 1979. Introduction. In *George Eliot's* Middlemarch *Notebooks: A Transcription*, ed. John C. Pratt and Victor A. Neufeldt, xvii–lii. Berkeley: University of California Press.

Reynolds, J. Russell. 1858. *On the Relation of Practical Medicine to Philosophical Method and Popular Opinion, Being the Annual Oration Delivered Before the North London Medical Society on February 10th, 1858*. https://wellcomelibrary.org/item/b22471807. Accessed 7 February 2020.

Richardson, Ruth. 1987. *Death, Dissection and the Destitute*. Chicago: University of Chicago Press.

Rothfield, Lawrence. 1992. *Vital Signs: Medical Realism in Nineteenth-Century Fiction*. Princeton: Princeton University Press.

Scudamore, Charles. 1826. *Observations on M. Laennec's Method of Forming a Diagnosis of the Diseases of the Chest by Means of the Stethoscope and of Percussion; and upon Some Points of the French Practice of Medicine*. London: Longman, Rees, Orme, Brown, and Green. https://wellcomelibrary.org/item/b21305882. Accessed 7 February 2020.

Shuttleworth, Sally. 1984. *George Eliot and Nineteenth-Century Science: The Make-Believe of a Beginning*. Cambridge: Cambridge University Press.

Sparks, Tabitha. 2009. *The Doctor in the Victorian Novel: Family Practices*. Aldershot: Ashgate.

Sutton, Thomas. 1813. *Tracts on Delirium Tremens, on Peritonitis, and on Some Other Inflammatory Diseases, and on the Gout*. London: Underwood. https://wellcomelibrary.org/item/b28148812. Accessed 7 February 2020.

Tambling, Jeremy. 1990. *Middlemarch*, Realism and the Birth of the Clinic. *ELH* 57: 939–60.

The New Annual Register of General Repository of History, Politics, and Literature, for the Year 1803. 1804. London: Robinson. https://books.google.de/books?id=7wgLAQAAMAAJ. Accessed 7 February 2020.

Trotter, Thomas. 1804. *An Essay, Medical, Philosophical, and Chemical, on Drunkenness, and Its Effects on the Human Body*, 2nd ed. London: Longman, Hurst, Reese, and Orme. https://wellcomelibrary.org/item/b21081670. Accessed 7 February 2020.

Vickers, Neil. 1997. Coleridge, Thomas Beddoes and Brunonian Medicine. *European Romantic Review* 8: 47–94.

Vitet, Louis. 1803. *Médecine expectante*, 6 vols. Lyon: Leroy. https://wellcomelibrary.org/item/b22042854. Accessed 7 February 2020.

Warhol, Robyn. 2002. The Rhetoric of Addiction: From Victorian Novels to AA. In *High Anxieties: Cultural Studies in Addiction*, ed. Janet Farrell Brodie and Mark Redfield, 97–108. Berkeley: University of California Press.

Warner, John Harley. 1997 [1986]. *The Therapeutic Perspective: Medical Practice, Knowledge, and Identity in America 1820–1885*. Princeton: Princeton University Press.

———. 2003 [1998]. *Against the Spirit of the System: The French Impulse in Nineteenth-Century American Medicine*. Baltimore: Johns Hopkins University Press.

Warning: This Drug May Kill You. Directed by Perri Peltz. 2017. Burbank: HBO.

Zieger, Susan Marjorie. 2008. *Inventing the Addict: Drugs, Race, and Sexuality in Nineteenth-Century British and American Literature*. Amherst: University of Massachusetts Press.

Index

A

absinthe, 12, 13, 16, 195–211

absinthism, 195–197, 200–204, 207, 210, 211

abuse. *See* drug abuse

Actor-Network-Theory (ANT), 11, 12, 156, 168, 169

addict, 61, 70, 71, 77, 79, 80, 83, 84, 86, 101, 110, 122, 123, 127, 135, 136, 139, 141, 149, 198, 247, 248, 259, 264. *See also* history of addiction

addiction, 1–3, 6, 8, 10–12, 48, 51–53, 56, 58, 61–64, 70, 71, 76–79, 82–86, 103–106, 120–123, 125, 127, 129, 131, 132, 136–139, 141, 142, 144, 149, 150, 168, 171–173, 176, 177, 186, 187, 195–201, 204–210, 221, 232, 242, 248, 256, 259, 264

agency, 3, 10, 11, 53, 56, 60, 62–64, 98, 135–138, 143, 146, 217, 219, 220, 225, 233

agentic. *See* agency

Akenside, Mark

Pleasures of the Imagination, 105

alcohol, 4, 6–10, 12, 51, 70–76, 78, 86, 94, 96, 97, 102, 103, 105–110, 119, 120, 123, 125, 128, 129, 141, 159, 171, 172, 195–202, 204–208, 210, 211, 243, 256, 273

alcoholic, 110

alcoholism, 72, 74, 75, 78, 107, 129, 195, 196, 198–201, 204, 207, 211. *See also* drunkenness

alkaloid(s), 95, 97, 100, 110, 123, 124, 168, 173, 265

Allanson, Mary, 36

Alston, Charles, 35

alternative medicine, 275

amber gris, 229

anaesthetic(s), 40, 247

N. Roxburgh and J. S. Henke (eds.), *Psychopharmacology in British Literature and Culture, 1780–1900*, Palgrave Studies in Literature, Science and Medicine,
https://doi.org/10.1007/978-3-030-53598-8

Analecta Pharmacognista, 95
anesthetic(s), 3
animal experiments, 203
Anstie, Francis Edmund, 51, 64
Stimulants and Narcotics, 279
ANT. *See* Actor-Network Theory
anthropocentric utilitarianism, 11, 156, 159, 160, 162
anthropocentrism, 11, 157, 162, 165, 168
antivenin, 205
anxiety, 64, 71, 77, 83, 142, 149, 168, 180, 234, 244, 250
aphrodisiac, 101, 196
Apothecaries Act, 270
apothecary/apothecaries, 27, 94, 96, 98, 103, 159, 231, 268–270, 280
appetite, 76, 77, 94, 129, 130, 166, 172, 176
Arabian jasmine, 102
Argand, Aimé, 100
Aristotle, 71
Organon, 276
Armstrong, Nancy, 52
assemblage, 11, 63, 137–140, 146, 148
asthenia, 199, 220, 232, 243, 266
Austen, Jane, 13, 14, 30, 217–236
"Kitty, or the Bower", 13, 217–219
Bower, 219, 220, 223
Kitty, 217, 219, 220, 222–224, 231, 235
Northanger Abbey, 13, 218, 224, 226, 235
Sense and Sensibility, 14, 218, 228, 231, 236
Austen-Leigh, James Edward, 218
Austin, Alfred, 77
autobiography, 80, 143
autonomy, 52, 63, 64, 136–138, 143, 150, 227
Awsiter, John, 160, 161, 165
An Essay on the Effects of Opium, 159

B

Babbage, Charles, 268
Bacon, Phanuel, 37, 96
balmy, 35
Barbauld, Anna Laetitia, 100
Barrett Browning, Elizabeth, 257, 258
Beddoes, Thomas, 4, 9, 100, 101, 111, 118, 120–122, 220, 221, 233, 243
Hygëia: or Essays Moral and Medical, 111
beer, 74, 76, 109, 198, 242. *See also* alcohol
Behn, Aphra, 34
Bennett, Jane, 137–140, 143, 149, 150
Bernard, Claude, 278
Berridge, Virginia, 28, 30, 48, 49, 77, 84, 85, 171, 174, 180, 181, 241, 242, 244, 245, 251, 266
beverage, 13, 74, 106, 108, 195–199, 202–204, 209
Bichat, Marie François Xavier, 264, 278
bioethics, 15
biography, 2
biomedical, 10, 137. *See also* medicine
biopolitics, 190. *See also* Foucault, Michel
biopower, 174. *See also* Foucault, Michel
bioprospecting, 227
Blackwell, Elizabeth, 225
blood stream, 12, 175, 179, 188, 189
Boccaccio, Giovanni, 27
body, 2–5, 7–12, 35, 36, 40, 55, 64, 71, 72, 82, 95, 100, 103, 118, 119, 121–128, 130, 137–140, 143–148, 150, 158–160, 162,

173–178, 186, 199, 200, 204, 206, 208, 221, 223, 232–234, 242, 244, 249, 268, 272, 276
bohemian, 208, 210, 264
botanical(s), 94–99, 162–165, 167–169, 219, 221, 223, 235
botanist(s), 95, 99, 165, 227
botany, 6, 13, 96, 156, 163, 219, 225–227, 232, 235
boundaries, 5, 6, 10, 15, 71, 72, 78, 82, 86, 98, 126, 129, 132, 174, 229, 231, 255
  class, 82, 85
  gender, 82
  racial, 85
  social, 82
Braddon, Mary Elizabeth
  Lady Audley's Secret, 79
Braidotti, Rosi, 148
  The Posthuman, 148
brain, 2, 4, 5, 6, 7, 9, 10, 24, 32, 34, 38, 119, 121, 122, 124, 125, 130, 147–149, 165–167, 174, 182, 185, 207, 222, 226, 234, 235. *See also* mind
Briggs, Asa, 266, 268, 269
Brooks, Maria, 247
Broussais, François-Joseph-Victor, 269, 272, 276
Brown, Charles Armitage, 31
Browning, Elizabeth Barrett, 48, 250, 257
Browning, Robert, 257
Brown, John, 4, 7, 9, 11, 25, 51, 101, 118, 138, 139, 160, 161, 165, 172, 174, 180, 199, 220, 221, 232–235, 243, 265, 266, 275
  Elementa Medicinae (The Elements of Medicine), 101, 120, 138, 160, 174, 180
Brown, Thomas, 30
Brunonian, 12, 25, 119, 120, 126, 127, 131, 139, 171–173, 176–180, 182, 189, 190, 234, 243. *See also* Brown, John
  Brunonianism, 3, 9, 31, 40, 51, 101, 138, 160, 173, 174, 181, 182
  Brunonian medicine, 51, 118, 220
  Brunonian theory, 25, 31, 33
Buchan, William, 37, 95, 222, 226, 232, 234, 264
  Domestic Medicine, 37, 95, 264
Budge, Gavin, 3, 4, 51, 55, 100, 101, 105, 108, 112, 126, 172–175, 179, 220, 232, 233, 243, 266
Burke, Kenneth, 162
Burton, Robert, 29, 34, 36
  Anatomy of Melancholy, 29, 34
Byron, Lord George Gordon, 32

C

cactus grandiflorus, 102
caffeine, 94, 96, 97, 104
cannabis, 9, 99, 202, 211. *See also* THC
capitalism, 11, 168, 172, 184, 187, 191
Cartesian dualism, 144. *See also* mind/body
Chadwick, Edwin, 272
Chartist, 182, 187
Charvet, Pierre-Alexandre, 28
  De l'action comparée de l'opium, 28
Chaucer, Geoffrey, 26, 27
Chawton, 220, 228, 229
chemist, 2, 101, 102, 270
chemistry, 30, 95, 98, 140, 147, 148, 163, 166
chimney, 9, 100, 101, 108
China, 49, 81, 83, 104, 183–187, 191, 192
chocolate, 2

Christen, Christopher, 247, 248
On the Nature of Opium, 247
Christian, 138, 178, 275
Christison, Robert, 266
civic good, 9, 94, 95, 100
class, 4, 12, 75, 76, 82, 85, 86, 175, 180, 182–184, 187–189, 192
lower-class, 74, 78, 79, 82, 85–87
middle-class, 48, 74, 79, 81, 82, 85, 161, 187
upper-class, 78, 79, 81, 85–87, 189
clinical gaze, 15, 278. *See also* Foucault, Michel
clinical medicine. *See* medicine
cocaine, 5
coffee, 2, 104
Coleridge, Samuel Taylor, 4, 14, 28, 38, 47, 48, 50, 100, 102, 103, 111, 121, 126, 160, 175, 180, 244–248, 250, 251, 253, 255, 256
"Eolian Harp", 111
"Kubla Khan", 28, 50, 58, 175, 245, 246, 253
Coleridge, Sara, 49, 249–251, 255–258
Collinson, Peter, 99
Collins, Wilkie, 49, 69, 78, 79, 82–84, 87, 171–173, 264
Armadale, 79, 84
Moonstone, The, 49, 84, 86, 172, 173
No Name, 79
Woman in White, The, 69, 82, 84, 86
colonial, 4, 6, 12, 94, 105, 110, 131, 162, 164, 173, 175, 176, 184, 185
colonial economy, 104
colonialism, 130, 190, 191
Combe, Andrew, 275
Combe, George, 275
community, 27, 158, 161, 178, 179, 188, 189
confessional literary autobiography, 157
confession, (genre of), 143, 158
Confessions. *See* De Quincey, Thomas
Constantinople, 241, 244
consumption, 10, 12, 14, 28, 31, 72, 78, 83, 85, 94, 98, 104–108, 110, 111, 119, 120, 122, 131, 135, 139, 141, 142, 144–146, 162–164, 166, 167, 169, 172, 173, 175, 177, 180, 186, 187, 189, 196–200, 202, 205, 206, 210, 211, 242, 245, 256, 264
Cooper, Thomas, 29
Corvisart, Jean-Nicolas, 278
Crabbe, George, 37, 264
Crumpe, Samuel, 24, 29, 34, 37, 265
An Inquiry into the Nature and Properties of Opium, 265
Cullen, William, 172, 218, 220–223, 226–228, 233, 235
Culpeper, Nicholas, 29
cultural practices, 16, 192
Curtis, William, 99
Flora Londinensis, 99
cyder, 109. *See also* alcohol

D

Darwin, Erasmus, 9, 16, 33, 93–96, 101, 102, 140, 221, 243, 244, 258
Botanic Garden, 95
Loves of the Plants, The, 9, 33, 93, 112
Phytologia, 94
Temple of Nature, The, 94
Zoonomia, 33, 94, 101, 102, 108, 109, 140, 141
Darwin, Mary, 107
Davy, Humphry, 2, 3, 100

Day, Horace B., 77
The Opium Habit, 78
de Beaumont, Jeanne-Marie LePrince, 109
Decadence, 197, 209
Decadent(s), 204, 207, 209, 210, 264
decriminalization, 1
Defoe, Daniel
Robinson Crusoe, 86
degeneration, 83, 84, 197, 201, 210
de Jussieu, Antoine-Laurent, 96
delirium tremens, 74, 273. *See also* alcoholism
democratic, 82, 85, 87, 94, 96
dependence, 8, 10, 56, 70–72, 76–78, 82, 83, 85, 86, 118, 122, 123, 128, 131, 132, 173, 177, 186, 222, 223, 256, 257
dependency, 94, 136, 173, 179, 256. *See also* addiction
depression, 50, 119, 168, 180, 182, 183, 186, 206, 251, 271
De Quincey, Thomas, 2, 7, 9–11, 14, 24, 25, 28, 30, 34, 37, 41, 47–50, 58, 80, 82, 83, 103, 118–123, 125–132, 135–137, 140, 142–150, 156–168, 175, 244–251, 255, 256, 258, 264, 273
Confessions of an English Opium-Eater, 24, 47, 80, 83, 103, 125, 135, 155
Suspira de Profundis, 58
Derrida, Jacques, 6
Disseminations, 6
Descola, Phillip, 165
Dickens, Charles, 49, 172, 264
All the Year Round, 172
dipsomania, 75
disease, 4, 15, 35, 51, 70, 72, 74, 78, 82, 172, 174, 196–198, 221, 222, 275, 276, 278–280. *See also* illness
doctor, 14, 15, 177, 178, 188, 224, 242, 243, 245, 248, 253, 256, 264, 266, 267, 278. *See also* medicine
Dodart, Denis, 95
Dodd, William, 39
domestic, 12, 14, 28, 79, 82, 85, 175–177, 182, 184, 185, 189, 190, 192, 218, 222, 224, 225, 228, 235, 245, 247, 249, 255, 259
domestic economy, 189, 190
domesticity, 12, 14, 176, 179, 189, 245
dose/doses, 9, 23, 29–32, 34, 38, 40, 41, 51, 77, 79, 84, 96, 102, 103, 108, 119, 120, 123, 146, 160, 198, 202, 221, 229, 242, 243, 250, 257, 258, 265, 273, 274, 276
dosing, 107, 222, 229
Douglas, Lord Alfred, 37, 207, 208, 210
Doyle, Arthur Conan, 5, 49, 264
dram drinking, 78, 85
dream, 48, 50, 54–58, 60–64
Dreyfus, Hubert, 158
drink, 12–14, 38, 72, 74–76, 98, 106, 127, 129, 188, 195, 197, 198, 208–211, 252
drowsiness, 7, 25, 26, 32, 36, 50. *See also* opium effects
drug abuse, 49, 127, 177, 181
drug action. *See* psychopharmacology
drug dependency, 136, 137. *See also* addiction
drug effect(s). *See* psychopharmacology
drug efficacy, 9, 94
drug experience, 144, 244

druggist, 101, 148, 164, 188, 189
drug(s), 1–15, 93–99, 101, 105, 107, 110–112
drug wars, 5
drunkenness, 72, 73, 75, 106, 109. *See also* alcoholism
Dryden, John, 26, 39, 48
dulling. *See* opium, opium effects

E

East India Company, 104, 161
ebriety. *See* drunkenness
economic, 5, 12, 94, 97, 104, 111, 173, 175, 176, 180–184, 187, 188, 190, 192, 204, 221
economy, 12, 58, 80, 94, 104, 109, 110, 112, 172, 175, 176, 181–185, 187, 189–192
Eddy, Mary Baker, 276
Edgeworth, Maria, 12, 31, 173–176, 178–180, 183, 189, 190, 192, 252
  Belinda, 12, 31, 173–177, 179, 181, 183, 189, 190, 192, 252
Edinburgh, 9, 101, 265, 274
Eliot, George (Mary Anne Evans), 15, 40, 250, 266–271, 273–275, 277, 278, 280
  Middlemarch, 15, 266–268, 270, 272–274, 277, 278
  "Quarry for *Middlemarch*", 266
emotion, 53, 55, 71, 98, 118, 120, 124, 130, 132, 139, 174, 179, 217, 219, 223, 224, 232
empire, 12, 100, 104, 175, 176, 184, 256
enhancer, 6
entanglement, 10, 120, 121, 124, 125, 144, 157, 168, 185
Epicurus, 105
epidemic, 132. *See also* opium epidemic
erasure (of knowledge), 218
escapism, 80, 81
ether, 97, 257, 258. *See also* gas/gasses
ethics, 5. *See also* bioethics
ethnocentrism, 191
eugenics, 197, 201, 202. *See also* degeneration
excitement, 8, 73, 174, 199, 243, 244, 248, 266
experiment(s), 2, 3, 9, 14, 24, 26, 84, 94, 95, 102, 106, 108, 144, 161, 165, 202, 211, 227, 230, 235. *See also* animal experiments

F

family, 36, 60, 79, 84, 102, 107, 177–179, 181, 185, 186, 189, 227–229, 256, 271
fashion, 13, 72, 75, 219, 221
fashionable, 13, 111, 218, 222–224, 231, 233
fatigue, 64, 180, 196
female community, 220, 228
female friendship, 220, 228, 236
female healers, 13, 218, 219, 224–227, 231
female patients, 236
femininity/feminine, 78, 79, 176, 178, 179, 224, 258, 259. *See also* gender roles
Fentanyl, 117, 120, 124, 125. *See also* opium
fever, 38, 167, 182, 185, 188, 189, 271, 272
finance(s), 12, 175, 180
fin de siècle, 16
flower(s), 34, 56–58, 60, 96, 106, 165, 166, 205, 209, 249, 251
folk remedy/remedies, 6, 220, 222, 227, 235, 236

form, 5, 7, 15, 32, 48, 49, 51–53, 55, 60, 81, 82, 97, 101, 120, 166, 168, 172, 177, 178, 199–201, 209, 211, 230, 245, 251, 257, 264, 269
Foucauldian. *See* Foucault, Michel
Foucault, Michel, 99, 158, 174, 178, 185, 272, 276–278
  Naissance de la Clinique (Birth of the Clinic), 277
Foxcroft, Louise, 2, 83, 84, 141, 142, 198, 200, 201, 205, 242, 244, 245, 248, 264, 266, 279
  The Making of the Addict, 141
France, 84, 196–198, 200, 202, 208, 209, 276, 277
French, 9, 13, 95, 96, 100, 196–198, 200, 201, 209, 210, 269, 276, 277, 279, 280
friendship, 224, 226
fruit, 57, 58, 96, 105, 106, 110, 166, 204–206
fungi, 94, 99
Fuseli, Henry, 48, 108
  Nightmare, 108

G
Galen, 94, 96, 274, 277. *See also* humourism
'Gallomaniac'. *See* Galen
garden, 103, 105, 111, 188, 219, 221, 223
Garth, Samuel, 33, 36, 38
gas/gasses, 3, 9, 94, 96, 98, 100–102, 204
Gaskell, Elizabeth, 12, 173, 174, 176, 181–188, 190–192
  Mary Barton, 12, 173, 174, 176, 181–185, 187–192
Gaskell, Peter, 72
gender, 12–14, 86, 175, 179, 187, 189, 192, 218, 223, 224, 228, 247, 250
gender roles, 259
genre, 4, 7, 8, 11, 24, 25, 40, 51–53, 59, 63, 69–76, 78–82, 87, 143, 158, 161, 172, 181
gothic novel, 81
Grant, Alexander, 34
  Observations on the Use of Opium, 34
Greg, W.R., 77
guilt, (narrative of), 157, 158
Guy's Hospital, 30, 31

H
habit, 10, 31, 47, 48, 50, 54, 76–78, 85, 104, 119–121, 128, 129, 131, 147, 173, 177, 221, 245, 248, 257
habit-forming, 8, 71, 76, 123, 131. *See also* addiction
habituation, 50, 61, 72, 73, 75, 82, 84, 86, 87, 94, 172, 177, 187
Hahnemann, Samuel, 276
Hall, Marshall, 268
  On Diagnosis, 268
Hall, Robert, 102
hallucinations, 32, 50, 74, 147, 148, 196, 197, 202, 207, 251
hallucinogenic, 13, 14, 50, 99, 167, 202
Hamilton, Robert, 29
Harlan, Aurelia Brooks, 59, 62
Harrison, William, 36, 37
  "In Praise of Laudanum", 37
Hazlitt, William, 38
healers, 14, 217–220, 226, 228, 231, 234–236
health, 4, 8, 9, 71, 73, 75, 76, 86, 93, 95, 97, 98, 100, 101, 103, 104, 108, 110–112, 120, 127,

139, 146, 149, 172, 174, 175, 177–181, 186, 189, 190, 198, 199, 220–222, 225, 227, 228, 231–236, 245, 255, 256. *See also* wellbeing
healthcare, 132, 219, 220, 228, 229, 235. *See also* medicine
herbal, 4, 13, 95, 96, 98, 218–220, 223–225, 227, 228, 231, 233, 236
herbalist, 29, 161, 188, 189, 275
herbal treatment, 234
hereditary, 107, 198, 199
heteronormative, 178
hibiscus, 102
Hichens, Robert, 13, 196, 204, 207–210
  Green Carnation, The, 204, 207, 209, 210
Hill, John, 35, 36
Hippocrates, 265, 276
history of addiction, 2. *See also* addiction
Holmes, Oliver Wendell, 277
homeopathy, 276
Homer, 25, 26, 28, 38
  Odyssey, 25, 28
homosexuality, 209
household book, 220, 228–230
Howard, John, 104
humanism, 62
humouralism, 3
humourist, 3
hunger, 126–129, 131, 132, 174, 176, 182, 185–187
Hutchinson, Sara, 30
hypodermic, 78
hysteria, 180, 234, 235

I

iatrochemical, 227
iatromechanical, 227
identity, 26, 71, 79, 80, 96, 97, 149, 150
illness, 6, 35, 118, 126–128, 131, 136, 174, 185, 188, 196–198, 200, 202, 203, 207, 211, 231, 235, 247, 251, 256, 258, 267. *See also* disease
Illustrations, 190
imperial, 104, 110, 173, 183, 184, 186, 187, 192
imperialism, 105, 184, 187, 191, 192
indeterminacy. *See* indeterminate
indeterminate, 13, 195, 196, 204, 210
India, 83, 84, 104, 118, 168, 176, 184, 187, 190
Indian Mutiny, 83
Indian Pink, 102
individual, 8, 12, 15, 49, 52, 53, 63–65, 71, 72, 80, 85, 86, 94, 95, 98, 105, 106, 109–112, 123, 136, 137, 139, 141, 143, 150, 176, 178, 181–183, 187, 189, 190, 192, 220, 222, 228, 232, 242–244, 254, 264, 279
industrialisation, 187
industrial novel, 173, 182, 183, 185, 187
inebriety, 78, 171
insanity, 15, 74, 75, 107, 109, 180, 199–201, 264
inspiration, 50, 58, 106, 195, 244, 254
intoxication, 33, 61, 62, 64, 74, 76, 94, 103, 104, 106, 108, 109, 163, 166, 221, 259
Invocation to Sleep, 37

J

Jacobin, 96, 100, 105, 111
James, Henry, 81, 267
Jenner, Edward, 268

Johnson, Samuel, 29, 35
Jones, John, 11, 24, 28, 33, 34, 37, 160, 161, 165, 265
  Mysteries of Opium Revealed, 33, 160, 265
juvenilia, 218

K
Kant, Immanuel, 121, 122
Keats, John, 7, 23–25, 30–40, 50, 111, 244
  Endymion, 38, 111
  "Ode to a Nightingale", 36, 38, 40, 50, 58
  "Sonnet to Sleep", 38
kitchen physic, 221, 223, 229

L
Laënnec, René-Théophile-Hyacinthe, 15, 276, 278
Lamb, Caroline, 37, 38
  'Invocation to Sleep', 37
Lancashire, 183, 184
Lancet, The, 200, 270
Landon, Letitia, 250–253
Latour, Bruno, 156, 161–163, 169
laudanum, 15, 29, 30, 35, 37, 79, 83, 84, 86, 101, 102, 104, 119–124, 127, 130, 159, 171, 176, 187, 230, 247, 250, 253, 254, 264–266
lavender drops. *See* lavender water
lavender water, 13, 217, 218, 221–226, 228, 229, 233–235
Lawrence, William, 280
leaves, 33, 60, 96, 100, 129, 186, 204
Leavis, Q.D., 267
Leigh, John, 24, 32, 37
Lewes, George Henry, 266, 267, 279
Lewis, William, 227, 229, 230, 234
  An Experimental History of the Materia Medica, 229
Linnaeus, Carl, 9, 26, 96, 99, 106
Lloyd, Martha, 221, 229–231
Logan, Maria, 35, 38, 49
London, 107, 129, 197, 201, 208, 220, 276
Louis, Pierre-Charles Alexandre, 277
Lucretius, 105
Ludlow, Fitz Hugh, 80
  Hasheesh Eater, The, 80
Lydgate, Tertius, 15, 266–280. *See also* Middlemarch
lyric, 8, 52, 53, 63
Lytton, Edward Robert Bulwar, 59, 61, 62
  King Poppy, 59–63

M
MacCulloch, John, 74
Macht, David, 3
Mackenzie, Henry, 254
  Man of Feeling, The, 254
Madden, Richard, 241, 242, 244
madness, 5, 39, 61, 62, 70, 75, 105, 110, 200
magazine, 70, 78, 121
Magnan, Dr Valentin, 199–203, 205, 210
Malthusianism. *See* Malthus, Thomas
Malthus, Thomas, 185
Manchester, 182–185, 187, 191
Mangan, James Clarence, 48
Mansel, Henry, 69, 75, 76, 172
marginalisation, 181, 218, 219, 223, 224, 226, 235
marigold, 102, 110
marriage, 39, 82, 102, 176, 182, 184, 186, 190, 224, 252
Martineau, Harriet, 173, 176, 181, 182, 190–192, 250, 256–258, 267

Illustrations of Political Economy, 173, 176, 190, 191
Letters on Mesmerism, 250
Life in the Sickroom, 250, 256, 257
masculinity/masculine, 78, 178, 181, 188. *See also* gender roles
master-slave metaphors, 165, 166
material culture, 223
materialism, 142
materialist, 3, 140, 150
materia medica, 14, 30, 35, 218, 221, 224, 227
matriarchy, 224, 228
mead, 106, 107
meadia plant, 106
Mead, Richard, 24, 29, 34, 106
médecine expectante, 15, 276
medical ethics, 15
medical humanities, 5
medical knowledge, 222
medical practitioner, 11, 94, 136, 139, 146, 160, 199, 274
medical profession, 13, 141, 142, 148, 188, 266, 267, 274, 278
medical professionalisation, 72, 269
medical science, 6, 135, 141, 142, 144. *See also* medicine
medication, 1, 4, 13, 30, 279
medicinal, 5, 13, 14, 29, 49, 51, 53, 72, 95, 104, 107, 126, 139, 196, 219–221, 225–227, 229, 232, 233, 235, 242, 245, 255, 256, 264, 270, 271, 276
medicine, 2, 4, 6, 13–15, 79, 177, 178, 180, 188–190, 218–231, 233, 235, 236. *See also* medical science
medical knowledge, 4, 140, 142, 173, 179, 188, 189, 223, 224, 231, 235, 268
melodrama, 70, 81, 182
mental disease. *See* disease
mental health. *See* health
metabolisation, 10, 119, 122, 124, 128, 129
metabolite, 9, 10, 124–126, 128, 130
metaphor, 39, 49, 63, 70–72, 78, 80, 82, 84, 86, 99, 118, 121, 123, 162, 163, 204, 205
Methodism, 179, 180
Miller, James, 75
Miller, J.H., 156
Milton, John, 26, 27, 130, 160
mind, 5, 7, 10, 12, 26, 27, 32, 33, 37, 39–41, 47, 50, 53, 55, 71, 72, 75, 103, 107–109, 118, 123–125, 128, 130, 144, 198, 199, 206, 208, 210. *See also* brain
mind/body, 142, 144, 145. *See also* Cartesian dualism
moderate consumption, 74
Monro, Alexander, 29
Montagu, Elizabeth, 263
Moonstone, 171
morality, 109, 111, 141, 144, 145, 199, 209, 248
moral responsibility, 146
moral(s), 8, 13, 56, 57, 59, 63, 64, 70, 71, 75, 76, 94, 109, 112, 118, 120, 130–132, 141, 143, 144, 146, 149, 150, 157–159, 198, 201, 208, 225, 242, 256, 259
morphine, 78, 123, 124, 125, 257–259, 265. *See also* opium
Morton, Thomas, 39
mushroom(s), 9, 98, 99, 102

N

narcotic(s), 8, 28, 51, 54–56, 64, 84, 97, 98, 102, 107, 141, 149, 174, 243, 259
narrative architecture, 156, 157

narrative(s), 8, 38, 48, 50–53, 59, 70, 81, 97, 126, 127, 129–132, 135, 143, 150, 156–159, 164, 177, 190, 218, 253, 273, 274
naturalism, 9, 96
neoclassical, 7, 23–26, 28, 29, 31, 32, 40, 41
nerve(s), 2–4, 13, 109, 118, 172, 199, 222, 224, 226, 230, 232–235
nervous disease. *See* disease
nervous illness. *See* illness
neurology, 5, 108
neuropathology, 199
neuroscience, 5
nitrous oxide, 2, 3, 100. *See also* gas/gasses
nonhuman, 137, 138, 143, 145, 148–150
novel, 8, 13, 15, 31, 34, 36, 48, 52, 53, 59, 63, 74, 76, 79, 80, 83–86, 166, 171–173, 175–186, 207, 209, 210, 228, 266, 280

O

Occidentalism, 191
oil, 96, 102, 196, 202, 229, 233, 272
O'Neill, Henrietta, 38, 49, 250, 252, 253, 255, 257
opiate epidemic. *See* opium epidemic
opiate(s), 9, 10, 14, 36, 38, 48, 50, 54, 55, 58, 64, 105, 117–120, 122–132, 205, 206, 242, 243, 245, 250–259, 265. *See also* opium
opioid, 1, 10, 117, 118, 121, 124, 130, 131. *See also* opium
opium
  opium dreams, 48, 50, 59, 246, 252
  opium-eater, 14, 77, 83, 84, 118, 119, 126, 128, 132, 135, 145, 146, 150, 157–159, 164, 167, 241, 242, 245, 259
  opium-eating, 9, 10, 47, 48, 76, 78, 81, 117–122, 125–132, 248, 259
  opium effects
    balm, 25, 35, 225, 254
    drowsiness, 7, 25, 26, 32, 36, 37, 50
    dulling, 7, 25, 32, 34, 37, 39–41
    forgetfulness, 7, 25, 26, 32, 38, 40, 50, 186
    indolence, 7, 25, 32, 34, 38, 40, 80
    insensible, 34
    lulling, 25, 32, 34, 37, 38
    oblivion, 25, 27, 32, 38, 40
    pain relief, 23, 25, 257
    sleep, 25–27, 29, 31–34, 36, 37, 39, 40
    stupefaction, 34
opium crisis. *See* opium epidemic
opium den, 49, 50, 264
opium epidemic, 1, 121, 131
Opium War(s), 4, 12, 29, 83, 175, 176, 182, 183, 186, 190, 192
orange lily, 102
"'Orestes': A Sonnet to Opium; Celebrating its Virtues", 35
oriental, 14, 51, 80, 82, 242, 244–248, 255
Orientalism, 175, 191
Otherness. *See* Other, the
Other, the, 72, 83, 86
overstimulation, 55, 172, 177, 179, 185, 232
Ovid, 27, 98, 109
  Metamorphoses, 98
OxyContin, 117, 120, 124

P

pain, 3, 10, 25, 26, 28, 32, 34, 35, 50, 54–56, 59, 64, 119, 120, 123, 124, 126–128, 130, 131, 157–160, 165, 167, 168, 257
Paine, Thomas, 100
painkillers, 14, 50, 117, 242, 243, 248, 265
painkilling, 123–125, 130
palliative, 64, 172, 273
papaver somniferum. *See* poppy
paraphernalia, 108, 178
Parkinson, John, 27, 34
patients, 30, 31, 34, 177, 179, 190, 205, 219, 220, 226, 230, 231, 233, 235, 243, 263, 264, 268–270, 272–275, 278–280
patriarchal, 14, 178, 228
Pechey, John, 29, 35
pedagogy, 99
Peisse, Louis, 277
penny fiction, 74, 82
Pepys, Samuel, 24
pestle, 101, 103
pharmaceutical, 117, 161, 168, 221, 227, 230, 236, 244, 274, 276, 279
pharmaceutic chemistry, 230
pharmacodynamics(s), 8, 119, 122, 123
pharmacognosy, 95, 96, 111
pharmacokinetic(s), 9, 10, 119, 121–123, 126, 131
Pharmacopeia Londinensis, 227
Pharmacopoeia Edinburgensis, 227
pharmacopoeias, 220, 227, 234
pharmacotherapy, 15, 266
pharmacy, 15, 229, 230
Pharmacy Act, 49, 85, 173, 241
pharmakon, 6, 10, 94, 95
physiology, 24, 95, 106, 109, 126, 165, 185, 222, 224, 278
pipe, 9, 100, 101, 103, 108
placebo, 123, 211
plant-based, 4, 13, 217, 225, 226, 228, 229, 231, 233–235
plant blindness, 164, 165
plant(s), 4, 6, 7, 9, 11, 12, 95–100, 102, 106, 109, 112, 141, 209
Plato, 6
  Phaedrus, 6
Platzner, Robert, 157
pleasure, 5, 6, 9, 34, 37, 53, 75, 97–99, 101, 104, 105, 110, 111, 126, 130, 148, 157–159, 164, 165, 167, 168, 242
poet, 23, 33, 37, 39, 47–59, 64, 248, 250, 277
poetic justice, 82
poetry, 7, 8, 25, 32–34, 36, 39, 40, 47–57, 59, 61–64
poison, 6, 10, 27, 74, 76, 97, 104–107, 110, 127, 138, 142, 160, 174, 200, 204–206, 209
political, 1, 4, 6, 11, 12, 15, 175, 176, 178, 182, 184, 187, 192
political agency, 187
Pollan, Michael, 156, 161, 163, 165, 166
polypharmacy, 15, 274
Pomet, Pierre
  Complete History of Drugs, 100
Pope, Alexander, 25, 26, 28, 39
  Dunciad, 39
poppy/poppies, 14, 25–27, 29, 33, 36–40, 56–64, 102, 103, 247–249, 251–253, 258
Porter, Roy, 3, 4, 218, 222–224, 226, 232, 256, 264, 275
posthumanism, 137, 138, 143, 145, 150
posthumanist, 10, 110, 136, 137, 143
post-opium, 32, 34, 40
poverty, 78

powder, 9, 94, 97, 101, 102
power, 11, 13, 14, 25, 31, 49, 53, 57, 58, 60, 74, 83, 85, 97, 98, 135, 136, 139, 141, 145–147, 149, 150, 174, 178, 186, 235, 252
practitioners, 15, 173, 199, 218–220, 222, 223, 227, 270, 275, 277, 280
prescribing, 5, 36, 94, 266, 272
prescription, 1, 4, 9, 13–15, 30, 96, 102, 117, 230, 231, 270, 275, 276
proto-feminist, 13, 14, 218, 223, 225, 229, 236
proto-pharmacological, 14, 225, 227, 235
psilocin, 9, 102
psilocybin, 94, 96
psychiatric medicine, 3
psychoactive, 1, 2, 8, 9, 11, 70, 71, 79, 82, 83, 85
psychoactive revolution, 2, 94
psychoanalysis, 147
psychology, 3, 201
psychopharmacological efficacy, 228, 229, 235
psychopharmacological knowledge, 3, 9, 11, 12, 122, 173, 183, 231, 236
psychopharmacology, 3–5, 7, 12, 15, 28, 119, 121, 125, 132, 150, 175–177, 181–183, 188–190, 192
  drug action, 3, 5, 10, 11, 15, 118, 119, 121–123, 125, 132
  drug effect(s), 3–5, 7, 8, 10, 11, 13–15, 30, 48, 53, 55, 64, 93, 119, 124, 129, 138, 142, 149, 155, 156, 160, 173, 185, 196–199, 204, 209–211, 242, 247, 251, 254
psychostimulants, 94, 100, 108
public good, 105. *See also* civic good

Q

Quarterly Journal of Foreign and British Medicine and Surgery, 247
Quincy, John, 229, 233, 234
Quinquet, Antoine, 100

R

race, 61, 84, 86, 175, 187, 189, 192, 201, 209
rationality, 143, 231
Reade, Charles, 78
realism, 73, 181
realist school. *See* realism
recreation. *See* recreational drug use
recreational drug use, 14, 15, 49, 242, 244
Reid, John, 243, 256
  Essays on Hypochondriacal and Other Nervous Affections, 243
remedy, 6, 10, 13, 29, 31, 35, 94, 110, 120, 139, 180, 189, 209, 217, 218, 221, 223, 225, 226, 228–230, 233, 234
responsibility, 60, 62, 136, 145, 150
restlessness, 206, 257–259
Reynolds, John Hamilton, 31
Reynolds, J. Russell, 280
rhetoric, 10, 56, 58, 70, 74, 76, 78, 83, 85, 86, 125, 128, 129, 131, 132, 155, 199
rhetorical, 70, 127, 129–132, 155–157
Richardson, Samuel, 36, 272
  Clarissa, 36
Robinson, Mary, 250–257
  "Invocation to Oberon", 250, 255
Romantic, 14, 48–50, 52, 63, 81, 94, 99, 102, 111, 121, 160, 167, 182, 241, 244, 252, 253

Romanticism, 111, 142
Rossetti, Christina, 13, 196, 204–206, 210
"Goblin Market", 204, 205
Royal College of Physicians, 24, 280
rum, 2
Rush, Benjamin, 277

S

Schelling, F.W.J., 97
Schiebinger, Londa, 6, 218, 219, 221, 224, 227, 230, 232
Schmidt, Johann A.
Lehrbuch der Materia Medica, 95
science, 3–6, 15, 93, 95, 111, 112, 141, 142, 144, 158, 199, 219, 267, 279
scientific knowledge. *See* science
Scudamore, Charles, 276, 277
sedative(s), 14, 30, 36, 51, 72, 98, 102, 139, 242–245, 248, 250, 252, 254–259, 265
seed(s), 96, 106, 110
seizure, 203, 206
self, 52, 61–64, 72, 83, 86, 112, 146–149
self-medication, 4, 13, 15, 107, 245, 264
self-dose, 223
self-healing, 219
sensationalism, 9, 69, 74, 85
sensation fiction, 8, 12, 70–83, 85, 86
sensation novel, 8, 49, 69–71, 73–83, 85–87, 172
sensibility, 3, 25, 34, 101, 130, 146, 222–224, 232
Seward, Anna, 37, 38, 49, 250–254, 257
"Sonnet XIII", 250, 252, 254
"To The Poppy", 37, 38, 250, 252, 254
Shakespeare, William, 26, 27, 160
Othello, 27
Shelley, Percy Bysshe, 111
"Ode to the West Wind", 111
Shelling, Friedrich Wilhelm Joseph von
First Projection of a System of Nature Philosophy, 97
slavery, 104, 105, 109, 110, 131
slaves, 105, 180
"Sleep and Poetry", 36, 40
sleeplessness, 180, 207, 250, 258, 259
smelling-bottle, 222, 223, 233
Smith, Charlotte, 250, 252
Desmond, 250
Smith, Thomas Southwood, 272
social good, 103, 267. *See also* civic good
soothing, 14, 25, 26, 37, 55, 104, 110, 223, 233, 235, 254, 255
Southey, Robert, 100
Spary, E.C., 94, 96, 97, 106
Eating the Enlightenment, 96
Spenser, Edmund, 26, 27
Faerie Queene, 27
spices, 2
spirits, 31, 74, 106, 107, 110, 129, 160, 188, 198, 220, 225, 229, 234, 250. *See also* alcohol
Steventon, 13, 220, 228, 235
sthenia, 199, 220
stimulant(s), 4, 8, 14, 25, 30, 51, 55, 76, 77, 85, 97, 98, 101, 105, 118, 131, 167, 171, 172, 174, 175, 177, 179–181, 221, 228, 233, 235, 242–245, 247, 248, 250, 251, 255, 256, 258, 259, 265, 275
stimulate/stimulation, 4, 25, 31, 50, 53, 76, 99–101, 104, 111, 112, 118, 138–140, 146, 172, 174, 180, 220, 242, 243, 249

stimulus/stimuli, 4, 31, 51, 118, 147, 150, 161, 173, 174, 177, 190, 191, 220, 221, 232, 242, 244
St James's Magazine, 73
stomach, 10, 119–122, 124, 126–131
sugar, 2, 83, 84, 105
sumptuary laws, 94
Sutton, Thomas, 273
Svenaeus, Fredric, 5, 15
Sydenham, Thomas, 29, 35, 265
sympathy, 104, 109, 111, 147, 183, 184, 190, 231, 232
syringe, 9, 103, 108
syrup, 29, 97, 275

T

Taylor, Alfred Swaine, 76
tea, 2, 4, 6, 77, 83, 97, 98, 103–105, 227
teetotalism, 74. *See also* temperance
temperance (movement), 74, 75, 78, 196, 198, 202, 208
Tennyson, Alfred, 8, 54–56, 63
  In Memoriam A.H.H., 54–56
  The Lotos-Eaters, 56, 63
terministic screens, 164
textile industry, 176, 182, 184, 186, 187
THC, 202. *See also* cannabis
Theriakis, 241, 242, 244, 256
thing-power, 137, 142, 143, 150
Thompson, Francis, 48, 56, 264
  "The Poppy", 56
Thomson, James, 34, 58
  Castle of Indolence, 34
  "To Our Ladies of Death", 58
Thomson, John, 274
  Account of the Life, Lectures, and Writings of William Cullen, 274
thujone, 12, 13, 195, 196, 199, 202–204, 207, 210, 211. *See also* wormwood
Tinsley's Magazine, 75
tobacco, 2, 4, 77, 84
tolerance, 77, 78, 173, 206, 265. *See also* addiction
torpor, 40, 136
toxicity, 106, 203
toxicological, 203
toxic(s), 93, 94, 97, 131, 202–205, 211
trade, 2, 5, 103, 105, 148, 181, 183, 185, 187, 190–192
transnational, 173, 175, 182, 190, 192
Trollope, Anthony, 267
  Doctor Thorne, 267
Trotter, Thomas, 72, 75, 141, 142, 199, 243, 273
  An Essay, Medical, Philosophical, and Chemical on Drunkenness, and its Effects on the Human Body View of the Nervous Temperament (Essay on Drunkenness), 72, 141, 243, 273
tuberculosis, 15, 31, 180
Turkey, 241

U

usquebaugh, 109
utilitarian, 11, 155–160, 168
utilitarianism, 158, 160, 162

V

Vesalius, Andreas, 278
Victorian, 7, 8, 13, 28, 40, 48, 50, 51, 53, 56, 59, 60, 64, 69–73, 75, 76, 82, 83, 85–87, 136, 141, 142, 144, 149, 174, 204, 206, 208–210

Virgil, 25, 26, 32, 36
Aeneid, 25, 26
vital materialism, 137, 142
vital materialist, 11, 136, 144, 150
Vitet, Louis, 15, 276, 277

W
Wakley, Thomas, 270
Walpole, Horace, 39
Watt, Ian, 52
Rise of the Novel, The, 52
Weber, F., 35
Wedgwood, Josiah, 102
wellbeing, 96, 98, 102, 104, 105, 106, 111. *See also* health
whiskey, 109, 128. *See also* alcohol
Whytt, Robert, 24
Wilberforce, Wilbur, 102, 103, 110, 264
Wilde, Oscar, 13, 49, 197, 207–210, 264
Wilson, Alexander Philip, 28, 29
An Experimental Essay, on the Manner in which Opium Acts, 28
Wilson, Elizabeth, 119, 122–125, 130
wine, 29, 74, 105, 107–110, 265. *See also* alcohol
withdrawal, 5, 10, 56, 61, 64, 73, 77, 119, 124, 125, 127, 128, 179, 206, 207, 243, 257, 265, 273. *See also* addiction
Wollstonecraft, Mary, 34, 36, 225, 226, 231
A Vindication of the Rights of Woman, 231
Mary: A Fiction, 34
women's enfranchisement, 236
women writers, 12, 173, 175, 189, 192
Wood, Ellen (Mrs Henry), 79, 82
East Lynne, 79, 82
Woolf, Virginia, 48
To the Lighthouse, 48
Wordsworth, William, 55, 58, 100, 101, 111, 160, 250
Lyrical Ballads, 55
wormwood, 6, 12, 196, 199, 200, 202, 205, 206. *See also* thujone
Wortley Montagu, Lady Mary, 30

Y
Young, George, 31, 242, 243
Treatise on Opium, 242

Z
Zieger, Susan, 2, 51, 52, 58, 74, 77–82, 122, 131, 173, 198, 200, 221, 242, 248, 256, 259, 264, 265, 274
Inventing the Addict, 259